Agnes E. Rupley, DVM, Dipl. ABVP–Avian
CONSULTING EDITOR

VETERINARY CLINICS OF NORTH AMERICA

Exotic Animal Practice

Virology

GUEST EDITOR
Cheryl B. Greenacre, DVM, Dipl. ABVP–Avian

January 2005 • Volume 8 • Number 1

SAUNDERS

An Imprint of Elsevier, Inc.
PHILADELPHIA   LONDON   TORONTO   MONTREAL   SYDNEY   TOKYO

W.B. SAUNDERS COMPANY
A Division of Elsevier Inc.

The Curtis Center • Independence Square West • Philadelphia, Pennsylvania 19106

http://www.vetexotic.theclinics.com

THE VETERINARY CLINICS OF NORTH AMERICA: Volume 8, Number 1
EXOTIC ANIMAL PRACTICE ISSN 1094-9194
January 2005 ISBN 1-4160-2833-1
Editor: John Vassallo

Reprints. For copies of 100 or more of articles in this publication, please contact the commercial Reprints Department, Elsevier Inc., 360 Park Avenue South, New York, New York 10010-1710. Tel: (212) 633-3813 Fax: (212) 633-3820, e-mail: reprints@elsevier.com

The ideas and opinions expressed in *The Veterinary Clinics of North America: Exotic Animal Practice* do not necessarily reflect those of the Publisher. The Publisher does not assume any responsibility for any injury and/or damage to persons or property arising out of or related to any use of the material contained in this periodical. The reader is advised to check the appropriate medical literature and the product information currently provided by the manufacturer of each drug to be administered to verify the dosage, the method and duration of administration, or contraindications. It is the responsibility of the treating physician or other health care professional, relying on independent experience and knowledge of the patient, to determine drug dosages and the best treatment for the patient. Mention of any product in this issue should not be construed as endorsement by the contributors, editors, or the Publisher of the product or manufacturers' claims.

The Veterinary Clinics of North America: Exotic Animal Practice (ISSN 1094-9194) is published in January, May, and September by W.B. Saunders Company; Corporate and editorial offices: The Curtis Center, Independence Square West, Philadelphia, PA 19106-3399. Accounting and circulation offices: 6277 Sea Harbor Drive, Orlando, FL 32887-4800. Subscription prices are $130.00 per year for US individuals, $215.00 per year for US institutions, $65.00 per year for US students and residents, $156.00 per year for Canadian individuals, $250.00 per year for Canadian institutions, $165.00 per year for international individuals, $250.00 per year for international institutions and $83.00 per year for Canadian and foreign students/residents. To receive student/resident rate, orders must be accompanied by name of affiliated institution, date of term, and the *signature* of program/residency coordinator on institution letterhead. Orders will be billed at individual rate until proof of status is received. Foreign air speed delivery is included in all *Clinics* subscription prices. All prices are subject to change without notice.

POSTMASTER: Send address changes to *The Veterinary Clinics of North America: Exotic Animal Practice*; W.B. Saunders Company, Periodicals Fulfillment, Orlando, FL 32887-4800. **Customer Service: 1-800-654-2452 (US). From outside of the US, call 1-407-345-1000.**

The Veterinary Clinics of North America: Exotic Animal Practice is covered in *Index Medicus*.

Printed in the United States of America.

GOAL STATEMENT

The goal of the *Veterinary Clinics of North America: Exotic Animal Practice* is to keep practicing veterinarians up to date with current clinical practice in exotic animal medicine by providing timely articles reviewing the state of the art in exotic animal care.

ACCREDITATION

The *Veterinary Clinics of North America: Exotic Animal Practice* will be offering continuing education credits, to be awarded by a school of veterinary medicine, contract pending.

The aforementioned school of veterinary medicine is a designated a provider of continuing veterinary education. Veterinarians participating in this learning activity may earn up to 6 credits per issue or a maximum of 18 credits per year. Credits awarded may not apply toward license renewal in all states. It is the responsibility of each participant to verify the requirements of their state licensing board.

Credit can be earned by reading the text material, taking the examination online at *http://www. theclinics.com/home/cme*, and completing the program evaluation. Each test question must be answered correctly; you will have the opportunity to retake any questions answered incorrectly. Following successful completion of the test and the program evaluation, you may print your certificate.

TO ENROLL

To enroll in the *Veterinary Clinics of North America: Exotic Animal Practice* Continuing Education program, call customer service at 1-800-654-2452 or sign up online at *http://www.theclinics.com/home/ cme*. The CME program is available to subscribers for an additional annual fee of $49.95.

CONSULTING EDITOR

AGNES E. RUPLEY, DVM, Diplomate, American Board of Veterinary Practitioners-Avian Practice; and Director and Chief Veterinarian, All Pets Medical & Laser Surgical Center, College Station Texas

GUEST EDITOR

CHERYL B. GREENACRE, DVM, Diplomate, American Board of Veterinary Practitioners-Avian Practice; Associate Professor, Department of Small Animal Clinical Sciences, College of Veterinary Medicine, University of Tennessee, Knoxville, Tennessee

CONTRIBUTORS

CRAIG FRANKLIN, DVM, PhD, Diplomate, American College of Laboratory Animal Medicine, Associate Professor, Research Animal Diagnostic Laboratory, Department of Veterinary Pathobiology, College of Veterinary Medicine, University of Missouri–Columbia, Columbia, Missouri

WILLIAM A. FRASER, BS, Biologist, Animal Disease Diagnostic Laboratory, Division of Animal Industry, Department of Agriculture and Consumer Services, Kissimmee, Florida

CHERYL B. GREENACRE, DVM, Diplomate, American Board of Veterinary Practitioners-Avian Practice; Associate Professor, Department of Small Animal Clinical Sciences, College of Veterinary Medicine, University of Tennessee, Knoxville, Tennessee

CHARLIE HSU, VMD, Research Fellow, Department of Veterinary Pathobiology, College of Veterinary Medicine, University of Missouri–Columbia, Columbia, Missouri

APRIL J. JOHNSON, DVM, Graduate Student, Department of Small Animal Clinical Services, College of Veterinary Medicine, University of Florida, Gainesville, Florida

CORINNA KASHUBA, DVM, Research Fellow, Department of Veterinary Pathobiology, College of Veterinary Medicine, University of Missouri–Columbia, Columbia, Missouri

MELISSA KENNEDY, DVM, PhD, Diplomate, American College of Veterinary Microbiologists, Assistant Professor, Department of Comparative Medicine, College of Veterinary Medicine, University of Tennessee, Knoxville, Tennessee

SCOTT W. KORTE, DVM, Research Fellow, Resident in Comparative Medicine, Department of Veterinary Pathobiology, College of Veterinary Medicine, University of Missouri–Columbia, Columbia, Missouri

ARIC P. KROGSTAD, DVM, Research Fellow, Department of Veterinary Pathobiology, College of Veterinary Medicine, University of Missouri–Columbia, Columbia, Missouri

ISABELLE LANGLOIS, DMV, Diplomate, American Board of Veterinary Practitioners-Avian Practice; Clinician, Zoological Medicine, Centre Hospitalier Universitaire Vétérinaire, Faculté de Médecine Vétérinaire, Université de Montréal, St-Hyacinthe, Québec, Canada

BARBARA D. PETTY, DVM, Assistant Professor, Aquatic Animal Health, Department of Large Animal Clinical Sciences, University of Florida, Gainesville, Florida

JANET E. SIMPSON, DVM, Research Fellow, Resident in Comparative Medicine, Department of Veterinary Pathobiology, College of Veterinary Medicine, University of Missouri–Columbia, Columbia, Missouri

JAMES F.X. WELLEHAN, DVM, MS, Zoological Medicine Service, College of Veterinary Medicine, University of Florida, Gainesville, Florida

CONTENTS

Subscription Information

ELSEVIER
SAUNDERS

VETERINARY
CLINICS
Exotic Animal Practice

Vet Clin Exot Anim 8 (2005) xi–xii

Preface

Virology

Cheryl B. Greenacre, DVM, DABVP-Avian
Guest Editor

To my fellow colleagues,

It has been a pleasure to bring this latest issue of *Veterinary Clinics of North America: Exotic Animal Practice* to you on the subject of virology. This issue is meant to provide practical and concise information for the clinical evaluation of rabbits, rodents, fish, reptiles, amphibians, birds, and ferrets. Also, there is an article that describes diagnostic testing modalities and another that outlines general virology. The contributing authors have worked diligently to develop outstanding articles with up-to-date information in a useful format. Did you know that frog skin can secrete a substance that has antiviral properties?

Our knowledge of viruses continues to evolve and our profession needs to stay abreast of the latest findings and apply that new knowledge to the betterment of our patients. I would like to encourage each one of you to contribute to this ever expanding field by being vigilant for new diseases, submitting specimens for necropsies or other diagnostic testing, and using the tools that are available to us. Many viruses that have been discovered in exotic animals would not have been described if this vigilance was not present in some of our colleagues. I would like to thank Dr. Branson Ritchie for encouraging this innate curiosity in me and many other colleagues. I also

doi:10.1016/j.cvex.2004.09.007   vetexotic.theclinics.com

would like to thank Dr. Ken Latimer for always being available to answer questions from those who are curious enough to ask.

Cheryl B. Greenacre, DVM, DABVP-Avian
Avian/Zoological Medicine
Department of Small Animal Clinical Sciences
College of Veterinary Medicine
University of Tennessee
2407 River Drive, C-247
Knoxville, TN 37996, USA

E-mail address: cgreenac@utk.edu

VETERINARY
CLINICS
Exotic Animal Practice

Vet Clin Exot Anim 8 (2005) 1–6

General concepts of virology

Melissa Kennedy, DVM, PhD, DACVIM[a],*,
Cheryl B. Greenacre, DVM, DABVP-Avian[b]

[a]*Department of Comparative Medicine, College of Veterinary Medicine,
University of Tennessee, 2407 River Drive, Knoxville, TN 37996, USA*
[b]*Department of Small Animal Clinical Sciences, College of Veterinary Medicine,
University of Tennessee, 2407 River Drive, C-247, Knoxville, TN 37996, USA*

A general review of concepts that relate to virology is provided for the practitioner to refresh and update the information retained since veterinary school. Topics covered include history and classification of viruses, virus pathogenesis, and various host and virus characteristics that influence pathogenesis.

History

With the advent of more sophisticated methods of evaluating virus DNA, it is becoming more apparent that many animals have coevolved with their respective viruses and that new ones have been introduced along the way. There is evidence that thousands of years ago, viruses (eg, small pox and polio) affected humans; these and other viruses were most likely present long before that [1]. Diseases that were caused by viruses were described long before the actual filterable agents were isolated and called viruses at the end of the nineteenth century. One important step in evaluating viruses was Koch's postulates. Dr. Koch stated that the following should be demonstrated before determining an agent as being the cause of a disease:

The agent must be present in every case of the disease.
The agent must be isolated and grown in vitro.
The disease must be reproduced when a pure culture is inoculated into
   a susceptible host.

---

* Corresponding author.
E-mail address: mkennedy2@utk.edu (M. Kennedy).

doi:10.1016/j.cvex.2004.09.010

The same agent must be recovered again from the experimentally infected host [2].

In 1936, nucleic acid was detected in viruses. In 1940, with the advent of the electron microscope, viruses were visualized for the first time and scientists began to realize their variety in shape and size. By 1949, viruses were being grown in eggs and cell and tissue culture [3].

Definition of a virus and terminology

Viruses range in size from a circovirus (14 nm) to a poxvirus (450 nm × 260 nm). Poxviruses are visible under light microscopy. A virion, or mature virus particle, is a simple structure that consists of DNA or RNA which is surrounded by a protein shell (capsid). A subunit of the capsid is called a capsomere; the polypeptide chains of the capsomere are referred to as chemical units. A nucleocapsid consists of nucleic acid and the protein shell together. Some viruses also may have a lipid envelope that is called a peplos; peplomeres are glycoprotein projections that may be present on the peplos [3].

The DNA viruses usually contain double-stranded DNA, whereas the RNA viruses usually contain single-stranded RNA. Examples of double-stranded DNA virus families include Herpesviridae, Poxviridae, Papovaviridae, and Adenoviridae. Single-stranded DNA virus families include Parvoviridae and Circoviridae. Examples of single-stranded RNA virus families include Paramyxoviridae, Rhabdoviridae, Coronaviridae, Togaviridae, Flaviviridae, and Orthomyxoviridae. Double-stranded RNA virus families include Reoviridae and Birnaviridae. Viruses replicate only in living cells and have no energy-yielding apparatus. Unlike bacteria, binary fission does not occur with viruses [4].

Classification of viruses

Viruses are classified by morphologic features and chemistry and are distinguished by properties of the genome, virion, and proteins and physical and biologic properties. Different properties that are observed in the genome include nucleic acid type—including distinguishing DNA versus RNA viruses—and genome size and sequence [4]. Different properties that are observed in the virion include capsid symmetry, presence of an envelope, and diameter. Different properties of the proteins include size, function, and amino acid sequence. Differences in physical properties mainly describe the stability of the virus under varied circumstances. Differences in replication strategy may be used to classify viruses by method of transcription, assembly, and release. The most varied differences between viruses is observed in the biological properties (eg, serology, host range, antigenicity, tissue tropism, transmission, vector, epidemiology).

Example of viral classification:

Order = Virales
Family = Viridae (eg, Papovaviridae)
Subfamily = Virinae (eg, Polyomavirinae)
Genus = Virus (eg, Polyomavirus)
Species = a number or animal species name (eg, passerine polyomavirus)

Virus pathogenesis

Virus pathogenesis is defined as the mechanism by which a virus replicates in the cell, and in doing so, injures a cell and produces disease. The capacity to produce disease, or the degree of disease, is termed "virulence." Possible outcomes of virus infection include cell death, persistence, latency, transformation, and nonproductive infection; however, most virus infections are asymptomatic [5].

Disease production commonly results from the virus doing what it needs to do to produce progeny. Virus replication involves several general steps:

The virus must first attach to a cell by way of a surface receptor, followed by entry or penetration into the cell

The virus uncoats within the cell and releases its nucleic acid

For most viruses, the next step is viral transcription to produce mRNA. The exception is the single-stranded, positive sense RNA viruses (eg, coronaviruses), whose genome is able to serve as mRNA and undergo immediate translation

For the remaining viruses, translation follows transcription

Replication of the genome is undertaken after initial protein production

After production of viral structural proteins and genome, the components are assembled and viruses are packaged into virions

Lastly, viruses exit the cell. For nonenveloped viruses, this usually occurs through cell lysis, but enveloped viruses may be released through membrane budding and leave the cell intact [6].

During this process, the virus essentially usurps the cellular machinery to carry out these various steps that lead to alteration of cellular function, and in many cases, cell death.

Pathogenesis can be defined at the level of the host. How does the virus enter the host? Where does it replicate? Does it disseminate to other tissues in the host? How is it shed and transmitted? Pathogenesis also can be defined at the level of the virus—What is the virus receptor? How does it enter the cell? How does the infection alter the cell? How is the virus released?

Host factors that influence pathogenesis

Many host factors influence pathogenesis, including genetic characteristics of the host and environmental influences. Genetic characteristics of the host include species, breed, organ, tissue susceptibility, and function at the cellular level (eg, cell receptor types and intracellular hospitality to the virus). Environmental influences of the host include immunity, age, stress, trauma, hormone levels, nutritional status, environmental conditions, and concurrent infection [5].

Virus factors that influence pathogenesis

Several virus factors influence pathogenesis, including cytocidal capabilities, cytotoxins, perturbation of cellular function, tissue tropism, and dose and route of inoculation. Some viruses produce proteins that are directly toxic to the cell. These proteins may be structural proteins or replication enzymes, but have lethal effects on the cell. In most cases, however, the death of the cell results from the virus "take over" of the cellular machinery for virus replication. Interference with cellular DNA, RNA, and protein synthesis may occur. The structures of the cell, including cell membrane and cytoskeleton, may be perturbed. Many viruses induce cell lysis for release from the cell, whereas others induce apoptosis. At a minimum, the function of the cell may be altered, and at most, the cell is destroyed. Lastly, cell death or injury may result from the host's immune response to the virus. This is due to destruction of infected cells and to so-called "bystander killing" of adjacent cells. Significant tissue damage can result in persistent infections where the antigen persists. Lesions may be due to production of immune complexes, cell-mediated destruction, and complement activation [5].

Cellular tropism is an important factor in determining the outcome of infection. It is defined as the specific cells and tissue in the host in which the virus replicates in a natural infection. It is determined by several factors, including the presence of the virus receptor (eg, cluster of differentiation 4 (CD4) expression for feline immunodeficiency virus (FIV) attachment), postpenetration events within the cell (eg, expression of appropriate cellular proteases for influenza), conditional parameters (eg, appropriate pH, temperature), and ability to cross a barrier (eg, blood–brain barrier). For example, viruses whose tissue tropism is the skin (eg, papilloma virus) are less virulent than those that target the central nervous system (eg, encephalitis viruses).

Course of infection

Viral infections are either acute or persistent. An acute infection generally has a rapid course with an incubation period of days to weeks and the virus clears the body within 2 to 3 weeks of disease onset. Persistent infections may last months to years and can be characterized as late complications of acute infection or latent, chronic, or slow infections. Persistent infections may be reactivated and cause acute episodes or cause late sequela to infections [5]. They may be associated with immunopathologic disease, may lead to neoplasia, and are important epidemiologically as a result of recurrence of or continual shedding.

Types of persistent infection include:

Late complications of acute infection—the virus persists, often in the brain (eg, "old dog encephalitis" due to canine distemper virus)

Latent infection—there is intermittent recrudescence of the disease and no virus is shed unless or until the clinical disease recurs (eg, herpesvirus in humans that causes chicken pox in a child and later can cause shingles in an adult)

Chronic infection—the virus is always demonstrable (eg, hepatitis B in humans when a small percentage of affected people become persistently infected and shed the virus continually which can then be associated with cirrhosis or carcinoma)

Slow infection—these viruses have a long incubation period followed by a slow progressive disease (eg, the spongioform encephalopathies such as "mad cow disease")

Many mechanisms lead to virus persistence (eg, the virus must not be overly cytologic and must avoid detection and elimination by the host immune system). Viruses avoid being overly cytologic by regulating gene expression or by producing less lytic variants. Antigenic variation allows evasion of the host immune system. Other mechanisms that a virus may use to persist include growing in protected sites, persisting in epithelial surfaces, inducing nonneutralizing antibodies or tolerance, or causing defective cell-mediated immunity [5].

Summary

A basic understanding of viruses and how they replicate and produce disease can aid in the management of virus infections. Parameters, such as clinical signs, sample and test selection, prognosis, and control, are implicit in this understanding. Information increases almost daily about known and emerging viruses; this impacts our ability to manage and control infections.

References

[1] Levine AJ. The origins of virology. In: Knipe DM, Howley PM, editors. Fields Virology, 4th edition. Philadelphia: Lippincott, Williams and Wilkins; 2001. p. 3–18.

[2] Murphy FA, Gibbs EPJ, Horzenik MC, Studdert MJ, editors. Viral Taxonomy and Nomenclature. In: Veterinary Virology, 3rd edition. San Diego: Academic Press; 1999. p. 23–42.

[3] Murphy FA, Gibbs EPJ, Horzenik MC, Studdert MJ, editors. The Nature of Viruses as Etiological Agents of Veterinary and Zoonotic Diseases. In: Veterinary Virology, 3rd edition. San Diego: Academic Press; 1999. p. 3–21.

[4] Condit RC. Principles of Virology. In: Knipe DM, Howley PM, editors. Fields Virology, 4th edition. Philadelphia: Lippincott, Williams and Wilkins; 2001. p. 19–51.
[5] Tyler KL, Nathanson N. Pathogenesis of Viral Infections. In: Knipe DM, Howley PM, editors. Fields Virology, 4th edition. Philadelphia: Lippincott, Williams and Wilkins; 2001. p. 199–243.
[6] Murphy FA, Gibbs EPJ, Horzenik MC, Studdert MJ, editors. Viral Replication. In: Veterinary Virology, 3rd edition. San Diego: Academic Press; 1999. p. 43–59.

VETERINARY
CLINICS
Exotic Animal Practice

Vet Clin Exot Anim 8 (2005) 7–26

Methodology in diagnostic virology

Melissa Kennedy, DVM, PhD, DACVM

*Department of Comparative Medicine, College of Veterinary Medicine, University of
Tennessee, 2407 River Drive, Knoxville, TN 37996, USA*

Definitive identification of viral infections can be challenging. Sometimes, history and clinical presentation are sufficient to narrow, if not solidify, the diagnosis. Laboratory testing often is required for confirmation of the identity of the infectious agent. Although commitments of time and money are involved, verification of the pathogen identity is important. Many diseases present with vague or similar symptoms and lesions. Depending upon the agent, aspects such as prognosis, treatment protocol, and control measures will be impacted by the diagnosis. In addition, identification of the infecting pathogen will aid surveillance and may be required by regulations that pertain to trade and transport. Finally, some agents have zoonotic potential [1,2].

Impediments to viral diagnoses exist. The variety of techniques and methods for detection of viral infection is nearly as diverse as the viruses themselves. Deciding which assay is most suitable in each case can be daunting. In addition, there is little, if any, standardization among diagnostic laboratories; this impacts the interpretation of results. The advent of newer molecular assays also offers challenges to the practitioner in terms of understanding the technique and assessing the results appropriately. Finally, proper interpretation of results can be difficult, especially with regard to serologic assays.

Diagnostic virology in its simplest form involves two avenues—detection of the virus itself and characterization of the host antibody response to the infecting virus. The assays that are involved vary in methodology, expense, availability, sensitivity, and specificity. This article provides an overview of some of the more common techniques that are used and their respective advantages and disadvantages.

First, it is important to remember that the results that are received from the various assays depend, in large part, on the quality of the specimen that

---

E-mail address: mkennedy2@utk.edu

was submitted. For antibody assays, this is straightforward, because only serum is required. It may be advantageous to ship these samples cooled in warm climes or during hotter time periods to prevent bacterial growth and antibody destruction. Proper sampling and shipment for virus detection may be more involved. For many assays (eg, antigen detection assays), the virus need not remain viable. In these situations, rapid shipment and cooling during transport are not required. For some assays, however, degradation of the virus during transport will necessitate cooling during shipment. This often is true for genetic detection that uses polymerase chain reaction. Many viral genomes, especially those of RNA viruses, are labile. For virus propagation, the virus must remain infectious from the moment it is collected until it reaches the laboratory. This requires rapid, cooled transport [1,2]. Some viruses are inactivated by freezing; therefore, as long as shipment is speedy, refrigeration/cold packs generally are sufficient [1]. Communication with the laboratory regarding shipping requirements will assist in the proper transport.

Virus detection

Detection of the virus that causes an infection uses one of several methods: The virus may be propagated in the laboratory and characterized; the virus may be visualized by electron microscopy; the viral proteins may be detected using specific antibody; the viral nucleic acid may be detected; or an activity of the virus, such as red blood cell agglutination, may be assayed. Identification of the virus that is associated with the disease is good evidence for identification of the etiologic agent. One must remember that the relationship between the agent that is detected and the disease is not always causal. Some of the more common assays are discussed below.

Virus isolation

For many viruses, virus isolation remains the gold standard for identification [1,3]. This assay requires collection of a sample that contains viable virus and propagation of the virus in a cell culture system. For some avian viruses, propagation in embryonated eggs is required [1]. After growth in the in vitro or in vivo system, the virus is characterized and identified.

Sample collection and transport are critical steps in successful virus isolation [1,2,4]. Usually, samples are collected from the disease lesions, using swabs, fluids, and tissue samples. Various secretions/excretions can be collected, depending on the mode of virus shedding. Viruses in which a viremic phase occurs following exposure may be isolated from blood; however, this most often is possible only in the early stages of infection. Collection of virus for virus isolation optimally is done in the acute phase. Later in infection or during chronic infection, the presence of host antibody may inhibit virus growth in the laboratory. The speed of shipment that is required depends on the lability of the virus; nonenveloped viruses tend to be hardier and less susceptible to inactivation than enveloped viruses [4]. Because the identity of the virus usually is not known, overnight shipment on cold pack is ideal. Certain information should be provided to the laboratory to facilitate the chances of successful propagation (eg, species, history, clinical presentation). Cell lines that are used vary with the suspected virus; any information that can assist the diagnostician in the appropriate selection of propagation medium is helpful.

After viral growth has been detected, identification of the virus is done. This can be accomplished in several ways. Some viruses produce characteristic cytopathic changes (eg, syncytia) in cell culture (Fig. 1). For viruses whose identity is completely unknown, electron microscopy allows visualization and classification of the virus into a family based on morphology. Specific identification usually is done through the use of virus-specific antibody in an assay, such as immunofluorescence. Alternatively, a virus activity may be exploited to identify the virus. For example, the virus of Newcastle disease, avian paramyxovirus–1, agglutinates chicken red blood cells (RBCs). Its presence in cell culture can be detected by agglutination of RBCs using the cell culture supernatant.

Virus isolation offers several advantages. It remains the gold standard for many viruses. It is more sensitive than the antigen detection assays (eg, immunofluorescence) because the virus is amplified effectively by propagation in the laboratory. Virus isolation is a nonspecific assay (ie, any virus present in the sample) may be propagated. This is in contrast to antigen identification, where the assay is specific. For example, submission of cells from a tracheal wash for the virus of infectious laryngotracheitis (herpesvirus) of birds will be negative if the agent is that of infectious bronchitis (coronavirus); however, if the sample is submitted for virus isolation, either virus may be isolated.

Virus isolation also has certain disadvantages. Generally, it is more time-consuming and expensive than most routine antigen detection assays [1]. It may take 1 to 2 weeks for the virus to be detectable and the increased material and labor increases the cost. Obviously, the virus must remain viable during shipment which can be a problem with certain labile viruses (eg, herpesviruses). Rapid shipment in refrigeration is required which may add to the cost. The presence of other components (eg, bacteria, toxins) can interfere with the ability of the virus to grow in the laboratory. For example, hepatic viruses can be difficult to isolate because of the presence of cell culture toxins in the liver tissue [2]. This also is true if neutralizing antibodies are present. Additionally, some viruses simply do not grow well in an in vitro system. This is true for some enteric viruses, for example.

Electron microscopy

Electron microscopy provides a means for visualizing virus from samples or laboratory propagation. Samples are homogenized, pelleted by

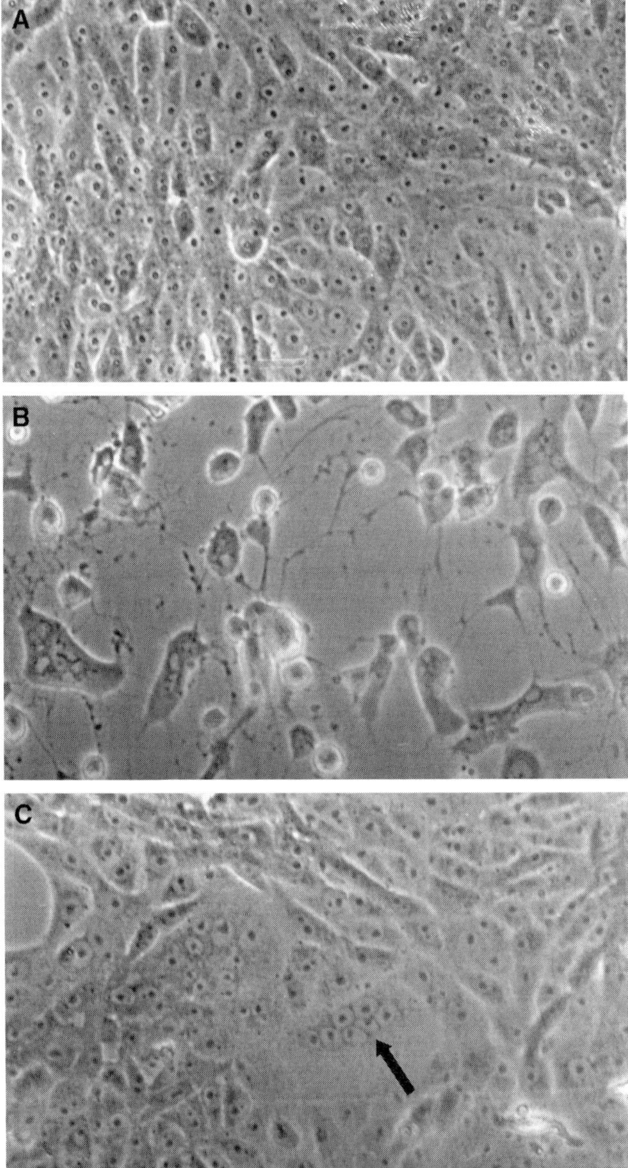

Fig. 1. Photomicrograph of cell culture exhibiting cytopathic effects. (*A*) Normal uninfected cells. (*B*) Cell death; note stringy appearance of cells, cells with vacuoles, and rounded detached cells. (*C*) Multi-nucleated syncytia (*arrow*).

centrifugation, and a small amount is mixed with a stain (eg, phosphotungstic acid) [1,5]. This is aerosolized onto a small grid and viewed by transmission electron microscopy. The virus, if present, is shadowed by the stain which gives an image that is similar to a negative photograph. The identification is based on morphology of the virion which allows classification into the virus family (Fig. 2). A more specific identification is provided by the addition of virus-specific antibodies to the mixture. This results in clumping of the virus [4,5].

Like virus isolation, electron microscopy is a nonspecific assay (ie, it looks for any virus that is present in the sample). For example, in diarrheic samples, any virus could be seen if present, whether parvovirus, coronavirus, or rotavirus, whereas an assay that is specific for one will miss the others. Where available, this assay is rapid and generally takes less than 1 day. In addition, the virus need not remain viable for detection and samples may be stored before shipment. Because feces is the most common sample that is submitted, storing and shipping under cooled conditions to keep down bacterial growth is recommended.

The equipment that is required for electron microscopy is extremely expensive as is the maintenance. Therefore, this service may not be widely

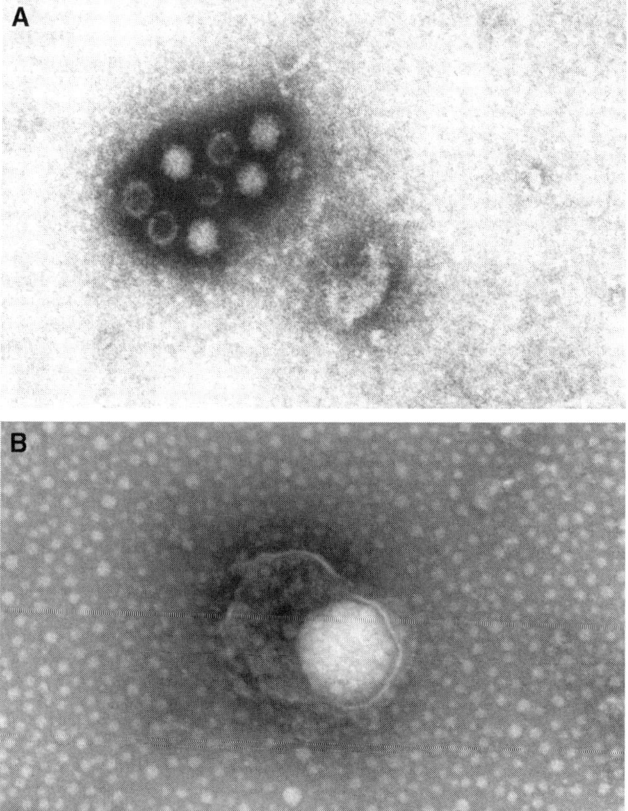

Fig. 2. Electron micrographs of two viruses (original magnification ×40,000). (*A*) Parvovirus. (*B*) Iridovirus.

available; it is primarily found at academic and research institutions. Also, expertise is required to identify the virus which also may impact availability. The fee for this service varies with the laboratory, but usually is reasonable, despite the expensive equipment. A drawback to electron microscopy is its relative lack of sensitivity. Generally speaking, approximately 10^6 particles per gram of sample are required for visualization [1,4,5]. More may be required for small viruses, such as parvovirus. Because of this, electron microscopy usually is restricted to testing of feces in cases of enteritis where virus concentration is significant in the acute phase of disease. For example, in diarrhea that is caused by parvovirus, the animal may be shedding more than 10^9 particles per gram. It also is used to identify viruses in cell culture, where the virus has been amplified through growth in the laboratory. It is too insensitive to detect virus in most tissue, fluid, or swab samples. These require propagation before detection. One exception to this may be poxviruses. Because of their large size, they may be seen in samples from lesions.

Antigen detection

Antigen detection assays involve the use of virus-specific antibody to detect or bind to the viral protein or antigen in the sample. The assays vary in the sample substrate, how the binding of antibody to antigen is visualized, sensitivity, and specificity. Availability of these assays for the various viruses depends largely on the availability of antibody to the pathogen. This can be a challenge for many viruses of exotic species because antibody to them often are not commercially feasible to develop. In these cases, one may find testing available at laboratories that are involved in research on the particular pathogen. We will consider several of the more common assays that are available.

Immunofluorescence assays (IFA)

Immunofluorescence assays (IFAs) are used in many diagnostic laboratories. They are fast and economical and can be done on a wide variety of samples. These include tissue impression smears or scrapings, cells pelleted from fluids, and cell cultures. Generally, slides are made by the practitioner and submitted to the laboratory. Neither fixation nor special storage is required before submission. The quality of the sample does affect the results. Because the assay involves detection of infected cells, the more cells that are on the slide, the more likely it is that infected cells will be identified. Most IFA assays use fluorochrome-labeled antibody, such as fluorescein (Fig. 3; direct) [1,3]. Therefore, it is important that before collection of ocular samples, fluorescein staining of the eye must be avoided. Also, purulent discharge will result in background fluorescence because of the autofluorescence of granules in white blood cells and should be removed before collection of the sample. Bacteria also may lead to background fluorescence.

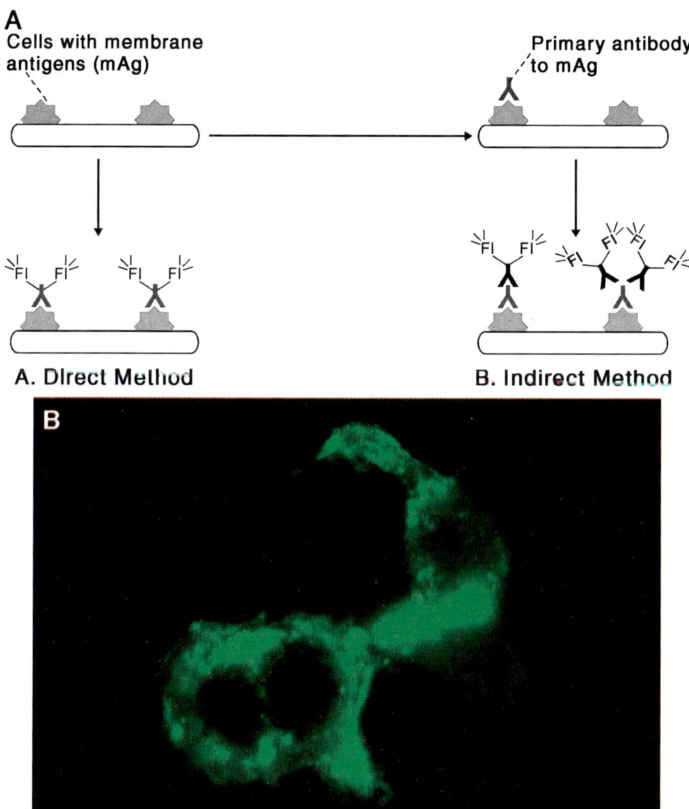

Fig. 3. (*A*) Direct and indirect immunofluorescence assays using fluorescein-labeled antibody. (*B*) Photomicrograph of immunofluorescence assay of virus-infected cells using fluorescein-labeled antibody.

These assays may be direct or indirect (see Fig. 3). In direct assays, the pathogen-specific antibody is labeled. In indirect assays, the virus-specific antibody is followed with an antibody to the first antibody (antiglobulin) that is labeled. This extra layer of antibody may increase the sensitivity of the antigen detection assays as a result of increased fluorescence relative to direct assays [4].

Immunofluorescence assays are intermediate in sensitivity. This sensitivity is even lower in chronic infections where the presence of host antibody can block the binding of the detecting antibody [6,7]. In general, a positive result "rules in" but a negative result does not "rule out." Also, because it is a specific assay, it does not detect the presence of other pathogens (eg, an IFA assay for felid herpesvirus on a conjunctival scraping will be negative if the pathogen involved is calicivirus). Finally, technical expertise is required to interpret IFA assays. The reader must be familiar with the pattern of fluorescence and have the ability to distinguish viral fluorescence from background or nonspecific fluorescence. Experience is thus required for proper interpretation of IFA assays.

ELISA and immunohistochemistry

ELISAs are available for some viral pathogens. These assays normally are provided as commercial kits and availability is dependent on marketability. Some available assays include feline leukemia virus, canine parvovirus, rotavirus, equine influenza virus, and bovine viral diarrhea virus. Most use pathogen-specific antibody for capture and detection as for IFA assays (Fig. 4B). The antibody is anchored in a solid substrate, such as a membrane or plastic well. The sample, usually in fluid form, is added. The initial sample collected may be fluid (eg, serum for feline leukemia virus) or the sample may be solubilized in a diluent using a swab (eg, feces for parvovirus). If the virus is present, it will be bound by the anchored antibody. This is followed by soluble pathogen-specific antibody with enzyme attached, which will bind the bound virus. Addition of the substrate for the enzyme leads to a color change. Washing is a critical step in many

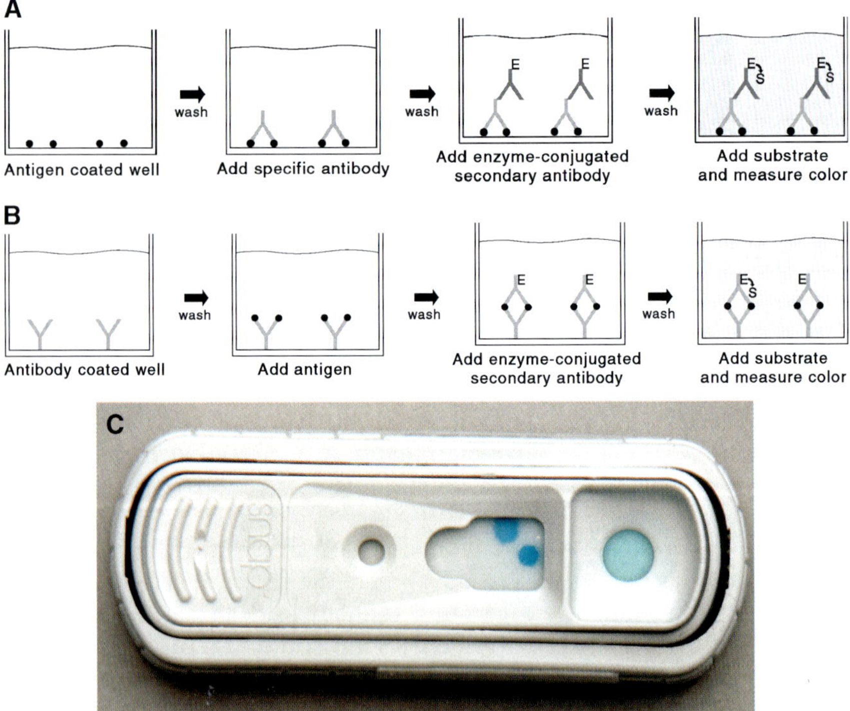

Fig. 4. (*A*) ELISA for antibody detection. (*B*) Diagram for ELISA for antigen detection. (*C*) Example of ELISA for Feline Immunodeficiency Virus (antibody) and Feline Leukemia Virus (antigen) (IDEXX Snap Test, Westbrook, Maine); FIV is positive (indicated by top blue spot).

ELISA assays to ensure unbound antibody is removed. Variations on this protocol may be found with different assays, but the underlying technique of pathogen-specific antibody binding the virus in the sample is similar in all (Fig. 4) [1,3,4].

ELISAs offer several advantages. Generally, they are rapid assays that require only minutes to 1 to 2 hours, at most, for completion and most are economical. They are simple and often can be done patient-side. Usually, ELISAs are more sensitive than IFAs. Because protein antigen is detected, the agent need not remain viable for identification.

ELISAs, like IFAs, are specific assays. Thus, a negative result will not provide information on the presence of other pathogens. Sensitivity of ELISA is usually good; however, negative results may occur early or late in infection when concentrations are low. Also, if the animal has large amounts of antibody present to the pathogen, these may bind and mask the agent and prevent binding by the assay antibody. False positive results may occur for a variety of reasons. The most common may be insufficient washing. Other factors can lead to false positive results. For example, some feline leukemia virus ELISAs use antibody to feline leukemia virus that is prepared in mice as the anchored and soluble antibody. Some cats have antibody to mouse antibody from hunting behavior; this can be bound by the anchored and soluble assay antibodies and lead to false positive results. Tests that use rabbit antibody can lead to the same result in felids with antirabbit antibody [8].

Immunohistochemistry (IHC) is similar to ELISA in that an enzyme-bound antibody is used to detect the virus in the sample (Fig. 5). The major difference is that the sample used is fixed tissue [9]. Sampling thus requires biopsy or necropsy. This allows identification of the virus associated with the microscopic lesion, if present.

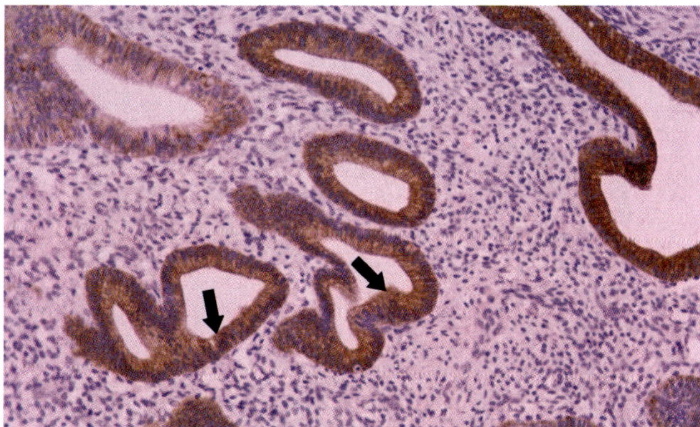

Fig. 5. Photomicrograph of uterine tissue labeled by immunohistochemistry (antibody to epithelial cell marker); note brown color labeling of epithelia (*arrows*). (Courtesy of R. Donnell, DVM, PhD, DACVP, Knoxville, Tennessee).

Usually, the sensitivity and specificity of IHC are good and it is the gold standard for some diseases, such as chronic wasting disease of cervids and feline infectious peritonitis [10,11]. Availability of this assay depends upon availability of the pathogen-specific antibody. Also, expertise in sample preparation and processing is required which also may impact availability.

Agglutination assays

Some pathogens have the ability to agglutinate RBCs. This assay usually is used to detect virus in cell culture [1,3]. Newcastle disease virus can be detected in the laboratory after propagation by detection of RBC agglutination. This assay is not done routinely on clinical samples for virus detection.

Agglutination of viruses also can be accomplished by the presence of pathogen-specific antibody in the assay; the immune complex formation can be visualized under certain conditions. One method is to attach the antibody to a solid substrate, such as a latex bead; addition of the sample with the antigen leads to clumping of the substrate which can be seen. The most common mechanism is through agar gel immunodiffusion (Fig. 6). In this instance, the immune complexes precipitate out in the agarose and form

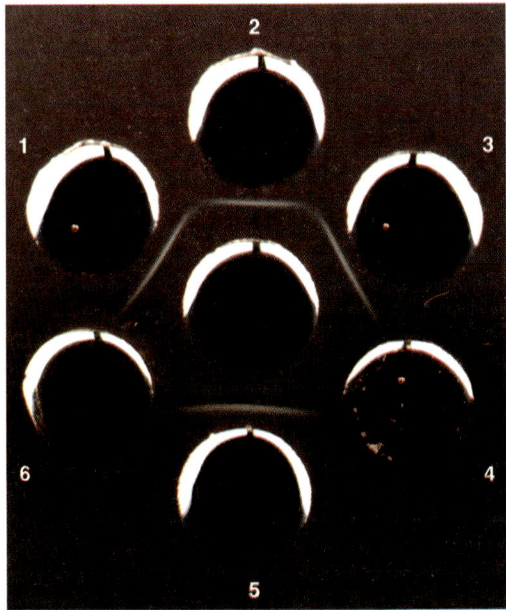

Fig. 6. Example of agar gel immunodiffusion for antibody to equine infectious anemia virus. Center well contains viral antigen; wells 1, 3, and 5 contain positive control sera; wells 2, 4, and 6 contain patient sera. Well 2 is positive, well 4 is weak positive, and well 6 is negative.

a precipitin band. This assay is low in sensitivity and is not used widely for detection of viral pathogens [1].

Genetic detection

The assays that were described above involve detection of whole virus or protein components of the virion. An alternative method is detection of the genetic material of the virus. A variety of techniques is available to accomplish this and are described below.

In situ hybridization

In situ hybridization (ISH) uses genetic probes to detect nucleic acid within the cell. Samples are as for IHC and usually involve fixed tissues, but pretreatment protocols differ from IHC. DNA or RNA probes are designed to hybridize to a portion of the pathogen genome or mRNA; their binding is detected through the labeling of the probe. The labels that are used can vary and include radioactive isotopes, fluorescent dyes, or chemiluminescent materials. The presence of the agent within the cell of the tissue sample can be visualized [9].

This technique allows visualization of the agent that is associated with the lesion and provides good evidence for identification of the etiologic agent of disease. There are several factors to consider with ISH to ensure appropriate interpretation of results. The sensitivity of the assay depends primarily upon two factors—the probe design and the abundance of the target. The probe design includes such parameters as the probe composition and length. The abundance of the target influences directly whether binding can be detected. Generally, for targets of low abundance, probes of increased length are required to increase the amount of label. The specificity also is impacted by the probe design. Is the genetic region that is targeted highly conserved or heterologous among isolates? Problems with nonspecific cross-hybridization also can be encountered.

The availability of this assay for particular pathogens depends on the presence of the required expertise and on the availability of sequence data for the virus of interest that is necessary for probe design. This technique may not be widely available for some pathogens of nondomestic species. Additionally, biopsy or postmortem tissue collection is required [12].

Polymerase chain reaction

Polymerase chain reaction (PCR) has become more common and accessible for detection of many pathogens. This technique involves the extraction of nucleic acid from the sample and amplification of a portion of the genetic material of the agent of interest. This amplification is

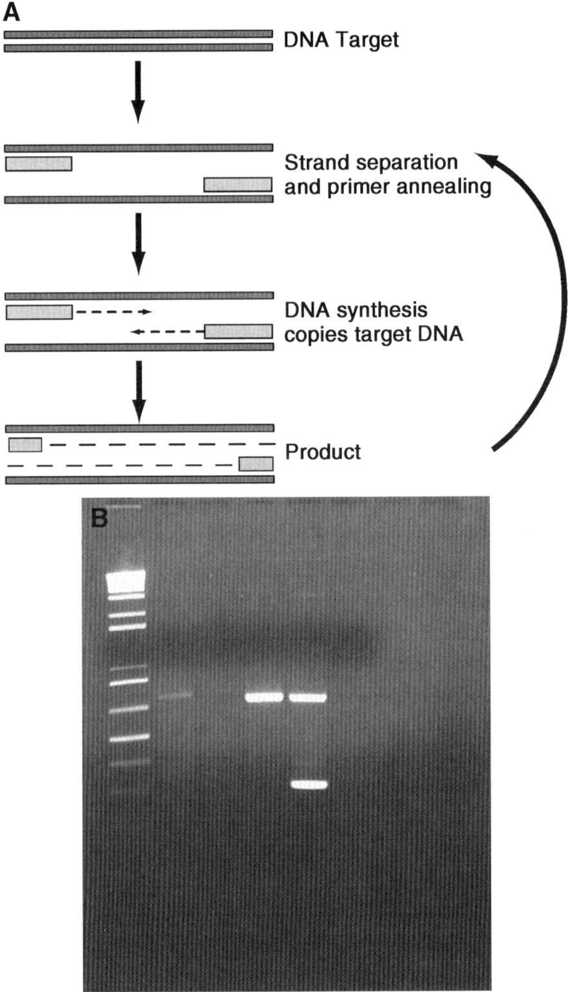

Fig. 7. (*A*) PCR. (*B*) Agarose gel electrophoresis analysis of DNA products. From left: lane 1, molecular weight markers; lane 2, weak positive; lane 3, negative; lane 4, positive; lane 5, positive control; lane 6, negative control.

accomplished by repeated cycles of DNA synthesis primed by small pieces of nucleic acid primers that are designed to target a specific genomic region (Fig. 7A). The resultant product is identified through gel electrophoresis (Fig. 7B), restriction enzyme digestion, probes, or nucleotide sequencing [3,4].

There are several variations to this assay. If the starting material targeted is RNA rather than DNA—such as with RNA viruses or assays that target mRNA of the agent—a reverse transcription reaction to convert the RNA to DNA is required before the actual amplification. Nested PCR involves two rounds of multiple amplification steps using two primer sets, one internal to the other. This allows significantly more amplification of the starting material. Quantitative PCR, also known as Real Time or TaqMan, allows quantification—either relative or absolute—of the starting material. Multi-plex PCR uses multiple primer pairs to allow detection of more than one pathogen in a single reaction [1].

One of the hallmarks of PCR is its exquisite sensitivity, especially with nested PCR, which can be a double-edged sword. It allows the detection of small amounts of the agent in the sample, which can be advantageous if low amounts are present [13–15]. Contamination or carry-over can lead to false positive results. In addition, subclinical or even latent infections (eg, herpesviruses) may be detected [16,17]. Real Time or quantitative PCR provides information on the amount of virus that is present in the sample. This can assist in distinguishing clinical from subclinical amounts of the virus in the sample.

False negative results also can occur and may be due to a variety of reasons. These may include degradation of the sample during transport, presence of PCR inhibitors in the sample, or insufficient genetic homology between the primers that are used and the target. This last reason can be a problem with viruses that undergo significant strain variation. In this case, primers should target highly conserved regions ideally [18]. For example, this has been an impediment to development of PCR assays that consistently work with all strains of FIV. RNA templates require reverse transcription before amplification; this step can be inefficient and may lead to false negative results [18].

Specificity of PCR also depends upon primer design, but generally, specificity of PCR is high. Depending upon the target, the assay can be family-, genus-, species-, or strain-specific.

Stringent controls and a significant level of expertise are required for accurate results using PCR. Contamination of samples can lead to false positive results. Nested PCR assays, because of their enhanced sensitivity and increased chance of sample contamination, are more likely to give false positive results than single rounds of PCR. Inquiries to the laboratory personnel regarding these parameters can provide useful information that can aid in interpretation of results. Often times, certain laboratories have more experience and expertise with certain pathogens. One laboratory that can test for all relevant pathogens is probably not feasible.

A variety of sample types can be used for PCR; it depends, in large part, on the pathogenesis of the agent. For example, viruses that are shed in feces can be tested for using fecal material; viruses that produce significant viremia may be tested for using serum or whole blood; viruses that are shed in feather dander (eg, psittacine beak and feather disease virus) may be detected in feathers [19]. Again, communication with the diagnostic laboratory will provide helpful information. Shipment of the sample usually requires refrigeration to prevent degradation of the genetic material of the agent. This is of particular concern with RNA targets because RNA is much more labile than DNA.

Serology

Evaluation of the host immunologic response involves detection of antibody to a specific pathogen. This provides, at the least, a historical perspective of the pathogens to which the animal has been exposed. By doing paired antibody assays—testing in the acute and convalescent phases of disease—the current infection status can be determined. A fourfold increase between acute and convalescent samples indicates active infection. Serology on herds or flocks can be used to establish the prevalence of the agent in the population [1].

The assays that are used to detect antibody are similar to ones for antigen detection—the antigen that is detected in serology is a pathogen-specific antibody. Unlike some antigen detection assays (eg, as virus isolation, electron microscopy), all serologic assays are specific (ie, they only detect antibody to a certain infectious agent). The assays for antibody detection can be divided into three basic categories [20]: (1) direct detection of the antibody, (2) detection of antibody-facilitated activity, and (3) detection of antibody-facilitated inhibition.

The first category involves binding of the animal's antibody to antigen anchored to a solid substrate (eg, membrane, slide). Detection of the bound antibody is done through anti-immunoglobulin (eg, antifeline IgG) that is labeled with a dye or enzyme [3,4]. These assays include immunofluorescence, ELISA, and western blot. This can be an obstacle in testing exotic species because there is limited availability of antiglobulin for most species of animals. Some that are available include antibody to immunoglobulin of ferrets, raccoons, llamas, psittacines, deer, bears, rabbits, monkeys, and several species of rodents. Testing for antibodies in other species may be available in some laboratories that are involved in research on the species [21]. If no antiglobulin is available, then testing is restricted to one of the other two categories.

The second category exploits the facilitation of an activity by the formation of the immune complex (patient's antibody; capture antigen, or virus); an activity occurs in response to formation of the immune complex formation. These may include complement fixation, agglutination, and precipitation. These assays detect IgG and IgM. Because IgM is an efficient activator of complement and enhances agglutination because of its pentameric structure, these assays are useful for detection of IgM [22,23]. The third category exploits the inhibition of an activity by the formation of an immune complex (ie, an activity that the virus normally is capable of effecting does not occur because of binding of the patient's antibody to the assay virus). These include hemagglutination inhibition (HI) and virus neutralization. These two assays measure antibody to specific surface epitopes of the pathogen, usually viral attachment proteins [3,4].

Serologic assays may be qualitative or quantitative [3]. The former determines only the presence of antibody, not the level. This can be useful for agents that cause life-long infection, such as the retroviruses; the presence of antibody would indicate infection. Quantitative assays determine the relative concentration of antibody by testing serial dilutions of the patient's serum. In this case, the titer is the highest dilution at which the presence of antibody is detected. Detection of seroconversion or a fourfold increase in titer between paired samples indicates active infection [1]. Paired titers can be circumvented in populations by screening a proportion of the animals. For example, testing exposed animals, clinically ill animals, and recovered animals can mimic acute and convalescent sampling. Basing a diagnosis on the magnitude of a single titer is unreliable; however, the laboratory may be able to provide some interpretation of a single titer.

The usefulness of serology depends, in part, on the virus and disease. If onset of clinical signs coincides with antibody production, active disease may be diagnosed. For some pathogens, disease occurs before antibody production; serology would be retrospective. Some pathogens cause disease months or years later, such as FIV; detection of antibodies is sufficient to diagnose infection [8].

Direct detection of the antibody

Most of these assays use antibody to the patient's immunoglobulin for detection. They vary in the capture antigen matrix (eg, slides, membrane, plastic wells) and label (eg, fluorescein, enzyme, chemiluminescent). Because these assays use species-specific antiglobulin, availability for exotic species is limited. Antiglobulin to antibody of domestic species may be used in some situations (eg, anticanine IgG in nondomestic canids like wolves). For those that are available, detection across species line is not always known. For example, in snakes, antiglobulin to IgY of boids does not detect IgY of elapids [21].

These assays vary in sensitivity and specificity. IFA assays are intermediate in sensitivity, whereas ELISAs and western blots are higher in sensitivity [22]. Generally, specificity depends upon the capture antigen that is used. For example, assays that use antigen that is shared with other agents that are nonpathogenic may detect antibody to these pathogens. Such an assay is antigen slides that contain whole *Rickettsia rickettsii* (Rocky Mountain spotted fever) organisms for detection of antibodies [24]. Because this agent shares antigenic epitopes with nonpathogenic rickettsial organisms, the presence of antibody does not confirm infection with the pathogenic agent [24]. Western blots have the highest specificity because antibody to the individual proteins of the agent, including antigens that are specific for the pathogenic species, is detected. For example, the western blot for Lyme disease can distinguish antibody to the vaccine from natural infection because of the difference in antibody profiles (vaccine: antibody to OSP A; natural infection: antibody to outer surface protein (OSP) C) [25].

IFA assay

The basis for immunofluorescence for serology is the same as for antigen detection. In this case, the antigen, usually virus-infected cells, is fixed to a glass slide. The anchored antigen "captures" the antibody that is present in the patient's serum (see Fig. 3; indirect). The bound antibody is detected with fluorescein-labeled antiglobulin. IFA assays are quantitative assays. The titer is reported as the highest dilution at which fluorescence is detected.

ELISA

The antigen in ELISAs is anchored to a membrane or plastic well. As with IFA assay the patient's antibody is captured and detected with antiglobulin, to which an enzyme is attached. Addition of the enzyme substrate leads to a color change (see Fig. 4). Most ELISAs are qualitative, rather than quantitative, assays with results reported as positive or negative. Most ELISAs are provided as kits by commercial suppliers and may be limited for diseases of exotic species. Some may be adaptable to exotic species, such as the FIV antibody ELISA, but one must recognize that these kits are not approved for use on nondomestic species and may not work consistently in all situations.

Western blot

Western blot is similar to ELISA. With western blot, however, the agent's proteins are separated on a gel before attachment to a membrane. In this way, the antibody response to specific proteins can be evaluated (Fig. 8) [22,23]. This can be helpful to confirm results of other assays. For example, FIV ELISA positive results are confirmed through western blot [8]. Because western blot detects antibody to individual pathogen proteins, it is more specific than other assays. False positive results that occur as a result of detection of antibody to nonspecific proteins are avoided. As with ELISAs, western blots do not quantitate antibody levels.

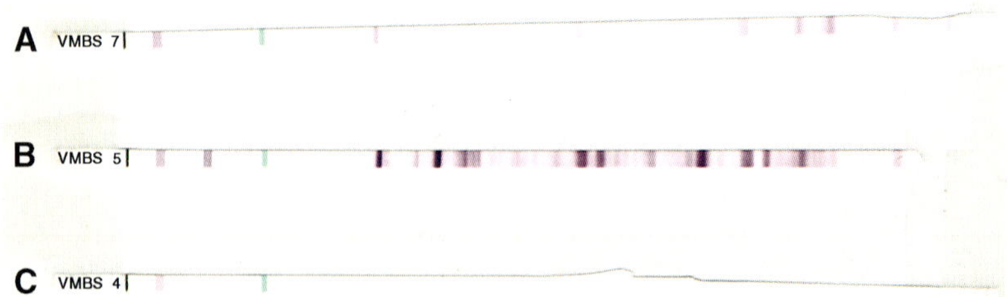

Fig. 8. Photograph of western blot for detection of antibody to Lyme disease agent. (*A*) Positive sample. (*B*) Positive control. (*C*) Negative control.

Detection of antibody-facilitated activity

Complement fixation

This assay is confined to detection of virus-specific antibodies. Although it has been invaluable in the past, it is not used commonly in most diagnostic laboratories. Formation of an immune complex by the binding of patient antibody to the virus leads to activation of the complement cascade. In the assay system, the incubation of virus with patient's serum is done with the addition of guinea pig complement. If antibody is present, the complement is activated and effectively exhausted. After the incubation, sheep RBCs, to which antibody has been attached, are added. These are, in effect, immune complexes. If the complement has been exhausted, the RBCs are not lysed by complement. If no antibody was present in the patient's serum, addition of the sensitized RBCs leads to complement activation and RBC lysis [1,3]. Most complement fixation assays quantitate the amount of antibody. Because IgM and IgG can activate complement, both will be detected. Generally, complement fixation is not used as widely as other antibody detection assays.

Agglutination/precipitation

Formation of immune complexes can be visualized through agglutination or precipitation of the complexes. In the former, the viral antigen may be attached to a substrate that can be visualized (eg, latex bead) or clumping of the organism itself may be seen [26,27]. In the latter, the formation of immune complexes leads to precipitation in an agarose medium at optimal concentrations, which can be visualized [28]. This is used commonly for detection of antibodies to caprine arthritis encephalitis virus and equine infectious anemia virus (see Fig. 6).

Detection of antibody-facilitated inhibition

Hemagglutination inhibition

This assay is used with certain viruses that have the ability to agglutinate RBCs. This assay is mediated by the surface proteins of the virus [1,3]. The patient's serum is incubated with the virus in a well. If present, binding of the patient's antibody to the virus will prevent the RBC agglutination by the virus. Therefore, the RBCs will pellet in the well. The highest dilution at which agglutination of the RBCs is inhibited is reported as the titer (Fig. 9). This assay measures antibody to epitopes on the surface protein of the virus. These proteins are under immunologic pressure for variability; thus, virus strains may vary in the antigenicity of these proteins [29]. Antibodies to highly conserved proteins (eg, polymerase or core proteins) are not detected

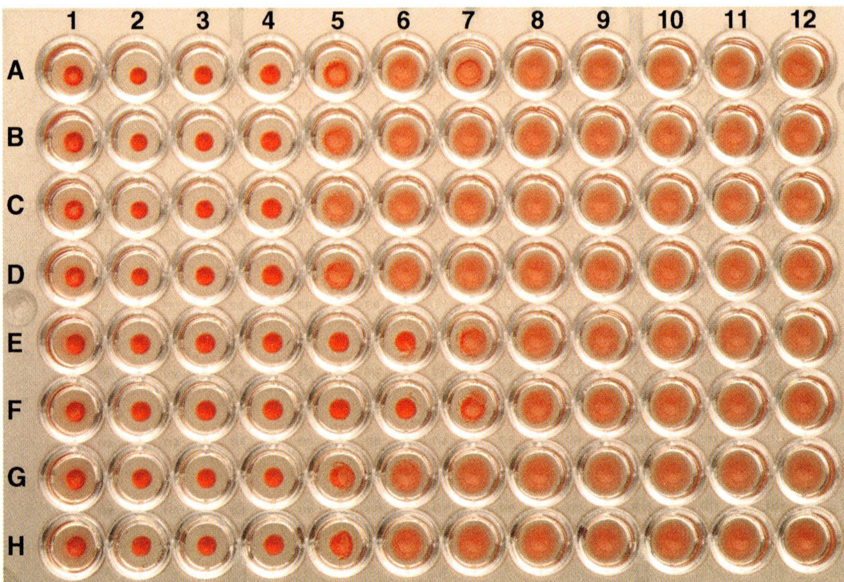

Fig. 9. Photograph of HI assay. Column 1: serum control with RBC and patient serum only. Columns 2–11: RBC, virus, patient sera in dilutions 1:20 to 1:5120. Column 12: virus control with RBC and virus only. Rows A–H: duplicates of patient serum. A/B: antibody titer of 1:80. C/D: antibody titer of 1:80. E/F: antibody titer of 1:320. G/H: antibody titer of 1:80.

by this assay. As a result, antibody that recognizes these specific epitopes on one virus strain may not recognize a different strain in this antigenic region. For example, HI is used often to measure antibodies in reptiles to OPV. We found that by using different isolates of ophidian paramyxovirus (OPV) in the assay, different antibody titers will result with the same serum because of variability of the epitopes among virus strains (unpublished data).

Serum virus neutralization

This assay is considered to be the most definitive serologic assay and may correlate with protection [1]. Binding of antibody to certain virus proteins inhibits the ability of the virus to infect a cell or neutralizes the virus. In this assay, the patient's serum is incubated with the virus, after which susceptible cells are added. If antibody is present, no infection will occur as evidenced by lack of cytopathic effects or detection of the virus in the cells by IFA assay. The absence of antibody allows the virus to infect the cells and this infection can be detected. The highest dilution at which no evidence of cellular infection (or infection of 50% of the cells) occurs is reported as the titer. As with HI, this assay commonly measures antibody to epitopes of surface proteins, and as a result, virus strain variation in these epitopes impacts the results.

Interpretation

Interpretation of serologic results can be a challenge. For quantitative assays, assigning an infection status to an animal based solely on the magnitude of a single titer is unreliable. Antibody may be absent or decreased in the acute stage of disease. Conversely, past infections with some agents lead to persistently elevated titers for significant periods of time [1]. In addition, titers vary depending upon the type of assay and capture antigen that are used. Finally, because laboratories vary in the methodology that is used, titers vary among them. Paired titers offer the most information, but are not always feasible. Communication with laboratory personnel can assist interpretation. For qualitative assays, a positive result does not correlate necessarily with active infection. Exceptions to this are those that cause lifelong infections (eg, FIV).

Summary

Selection of proper assays and appropriate interpretation of results can be a challenge to the veterinary clinician. Assays vary in methodology, sensitivity, and specificity. These assays can be invaluable in attaining the correct diagnosis, but a clear understanding of the assay and the results is essential. To this end, communication with the laboratory personnel is crucial. Optimal sample selection, shipping recommendations, assay selection, and interpretation should be discussed with the laboratory staff.

References

[1] Quinn PJ. Laboratory diagnosis of viral infections. In: Quinn PJ, Markey BK, Carter ME, et al, editors. Veterinary microbiology and microbial disease. Malden (MA): Blackwell Science; 2002. p. 305–12.

[2] Heuschele WP, Castro AE. In: Castro AE, Heuschele WP, editors. Veterinary diagnostic virology. St. Louis (MO): Mosby Year Book; 1992. p. 2–8.

[3] Burleson FG, Chambers TM, Wiedbrauk DL. Virology: a laboratory manual. San Diego (CA): Academic Press, Inc.; 1992.

[4] Murphy FA, Gibbs EPJ, Horzinek MC, et al. In: Veterinary virology. 3rd edition. San Diego (CA): Academic Press; 1999. p. 193–224.

[5] McClean DM, Wong KK. Same-day diagnosis of human virus infections. Boca Raton (FL): CRC Press, Inc.; 1984.

[6] Maggs DJ, Lappin MR, Reif JS, et al. Evaluation of serologic and viral detection methods for diagnosing feline herpesvirus-1 infection in cats with acute respiratory tract or chronic ocular disease. J Am Vet Med Assoc 1999;214(4):502–7.

[7] Stiles J, McDermott M, Willis M, et al. Comparison of nested polymerase chain reaction, virus isolation, and fluorescent antibody testing for identifying feline herpesvirus in cats with conjunctivitis. Am J Vet Res 1997;58:804–7.

[8] Barr MC. FIV, FeLV, and FIPV: interpretation and misinterpretation of serological test results. Semin Vet Med Surg (Small Anim) 1996;11(3):144–53.

[9] Harvey R, Farmilo AJ, Stead R. In: Boenisch, editor. Immunochemical staining methods. 3rd edition. Carpinteria (CA): DakoCytomation; 2001. p. 18–22, 41–3.

[10] Addie DD, Paltrinieri S, Pedersen NC. Recommendations from workshops of the Second International Feline Coronavirus/Feline Infectious Peritonitis Symposium. J Feline Med Surg 2004;125–30.

[11] Salman MD. Chronic wasting disease in deer and elk: scientific facts and findings. J Vet Med Sci 2003;65(7):761–8.

[12] Garcia AP, Latimer KS, Niagro FD, et al. Diagnosis of polyomavirus-induced hepatic necrosis in psittacine birds using DNA probes. J Vet Diagn Invest 1994;6:308–14.

[13] McElnea CL, Cross GM: Methods of detection of *Chlamydia psittaci* in domesticated and wild birds. Aust Vet J 77(8):516–21.

[14] Horner GW, Tham KM, Orr D, et al. Comparison of an antigen capture enzyme-linked assay with reverse transcription-polymerase chain reaction and cell culture immunoperoxidase tests for the diagnosis of ruminant pestivirus infections. Vet Microbiol 1995;43:75–84.

[15] Sandvik T. Laboratory diagnostic investigations for bovine viral diarrhea virus infections in cattle. Vet Microbiol 1999;64:123–34.

[16] Burgesser KM, Hotaling S, Schiebel A, et al. Comparison of PCR, virus isolation, and indirect fluorescent antibody staining in the detection of naturally occurring feline herpesvirus infections. J Vet Diagn Invest 1999;11:122–6.

[17] Weigler BJ, Babineau CA, Sherry B, et al. High sensitivity polymerase chain reaction assay for active and latent feline herpesvirus-1 infections in domestic cats. Vet Rec 1997;140:335–8.

[18] Sellon RK. Update on molecular techniques for diagnostic testing of infectious disease. Vet Clin Small Anim 2003;33:677–93.

[19] Ypelaar I, Bassami MR, Wilcox GE, et al. A universal polymerase chain reaction for the detection of psittacine beak and feather disease virus. Vet Microbiol 1999;68:141–8.

[20] Storch GA. In: Knipe DM, Howley PM, editors. Fields virology. 4th edition. Philadelphia: Lippincott, Williams and Wilkins; 2001. p. 493–531.

[21] Kania SA, Kennedy MA, Nowlin N, et al. Development of an enzyme linked immunosorbent assay (ELISA) for the diagnosis of ophidian paramyxovirus (OPV). Infect Dis Rev 2000;2(4):213–7.

[22] Kuby J. In: Immunology. 3rd edition. New York: WH Freeman and Company; 1997.

[23] Tizard IR. In: Veterinary immunology: an introduction. 6th edition. Philadelphia: WB Saunders; 2000.

[24] Greene CE, Breitschwerdt EB. In: Greene CE, editor. Infectious diseases of the dog and cat. 2nd edition. Philadelphia: WB Saunders Company; 1998. p. 155–62.

[25] Littman MP. Canine borreliosis. Vet Clin North Am 2003;33(4):827–62.

[26] Grimes JE, Tully TN, Arizmendi F, et al. Elementary body agglutination for rapidly demonstrating chlamydial agglutinins in avian serum with emphasis of testing cockatiels. Avian Dis 1994;38:822–31.

[27] Mateu-de-Antonio EM, Casal MMJ. Comparison of serologic tests used in canine brucellosis diagnosis. J Vet Diagn Invest 1994;6:257–9.

[28] Knowles DP. Laboratory diagnostic tests for retrovirus infections of small ruminants. Vet Clin North Am: Food Anim Pract 1997;13:1–11.

[29] Prince HE, Leber AL. Comparison of complement fixation and hemagglutination inhibition assays for detecting antibody responses following influenza virus vaccination. Clin Diagn Lab Immunol 2003;10(3):481–2.

ELSEVIER
SAUNDERS

Vet Clin Exot Anim 8 (2005) 27–52

VETERINARY
CLINICS
Exotic Animal Practice

Reptile virology

James F.X. Wellehan, DVM, MS*,
April J. Johnson, DVM

College of Veterinary Medicine, University of Florida, Gainesville, FL 32610, USA

The body of knowledge, on reptile virology is expanding rapidly. Reptiles are a diverse group of species; much work remains to be done to elucidate the viruses of reptiles. In addition to frank viral diseases, viral infections may play a role in the establishment of bacterial, fungal, and parasitic diseases and may be an underlying cause of neoplasia. It is important for the reptile clinician to recognize disease syndromes and pathology that are consistent with viral etiology, to understand viral diagnostic testing, and to have good communication with an experienced reptile pathologist and a good viral diagnostic laboratory.

Diagnostic tests

Diagnostic tests for the viruses of reptiles may be divided into two categories—testing for immune response to virus and testing for the presence of virus. It is important for the clinician to be aware of the information that is given by the test and to interpret results appropriately.

Tests that look for the presence of immune response to virus include serum neutralization testing, ELISAs, hemagglutination inhibition (HI), agarose gel immunodiffusion testing, and T-cell proliferation assays. It is important to consider what aspect of immune response is being tested. T-cell proliferation assays test for the presence of a cellular immune response, whereas the others test for the presence of a humoral immune response. A humoral immune response primarily produces antibodies against pathogens; a cellular immune response primarily uses cytotoxic T cells to trigger self-destruction of host cells that are infected with pathogens. Typically, a cellular immune response is most protective for viruses. Cellular and

---

* Corresponding author.
E-mail address: wellehanj@mail.vetmed.ufl.edu (J.F.X. Wellehan).

1094-9194/05/$ - see front matter © 2005 Elsevier Inc. All rights reserved.
doi:10.1016/j.cvex.2004.09.006

humoral immune responses often—but not always—are found together. An animal may have a cellular immune response without a significant humoral immune response or vice versa.

It also is important to consider whether the patient has had time to develop an immune response to a pathogen. This may take several weeks in reptiles [1]. In acute infections, a specific immune response may not be present. There are large seasonal and temperature effects on the immune response of reptiles, as well as species differences [2–4]. Other factors, including reproductive status and exogenous corticosteroid administration, also may affect the development of a measurable immune response [5,6].

Virus isolation or polymerase chain reaction (PCR) gives information regarding the presence of virus. These tests require virus in the sample submitted. Choice of sample for these tests is highly important. For example, in a patient that has a viral infection that is not viremic at the time of sample collection, test results on blood will be negative. This does not mean that there is not virus elsewhere in the patient. For virus isolation, the virus needs to be viable; the sample needs to be collected and stored in a manner that does not inactivate the virus. PCR testing requires intact nucleic acid, but does not require live virus.

PCR testing is highly sensitive which makes it susceptible to false positives as a result of contamination. A PCR involves the use of primers that are specific for a nucleic acid (DNA or RNA) sequence and repeated use of an enzyme to synthesize copies of the nucleic acid sequence between them. Primers may be designed to be highly specific for an organism or to amplify nucleic acid from any organism in a given group. A PCR that uses primers that are designed to amplify nucleic acid from multiple organisms is called a consensus PCR. After nucleic acid has been amplified by PCR, the identity of the product needs to be validated. The least specific method of validating the PCR product is gel electrophoresis, which shows that the product is the correct length. To increase the specificity, the product first may be digested by restriction enzymes, which recognize short sequences and cut them. This shows that the PCR product contains restriction enzyme sequences that are found within the expected product. A more specific method of validating the identity of the PCR product is to hybridize the product with a labeled nucleic acid probe from the expected product. The most specific method is to sequence the PCR product. Occasionally, a good diagnostic laboratory will sequence PCR products to ensure quality control. This should be available as a validation to any clinician who submits a sample for PCR, for an additional cost. Any consensus PCR product should be identified by DNA sequencing.

If the presence of virus is identified by virus isolation, electron microscopy, or nonspecies-specific DNA in situ hybridization, it is strongly recommended that the virus be identified further. Identification of viral types and species provides useful diagnostic, prognostic, and epidemiologic information for the clinician. Significant clinical differences may exist between viruses in the same family from the same host species and between different host species with the same virus. In primates, this is illustrated by the clinical difference between infection with human herpesvirus 1 (herpes simplex), which causes cold sores in humans, and the closely related cercopithecine herpesvirus 1 (herpes B), which causes cold sores in macaques and a fatal encephalitis in humans. In gibbons or marmosets, human herpesvirus 1 causes a fatal encephalitis.

Regardless of the type of diagnostic testing for virus, this information needs to be correlated with clinical and histologic findings in the animal. Neither the presence of an immune response nor the presence of virus necessarily indicates the presence of viral disease. Virus infection may occur without clinical disease and the presence of an immune response does not mean that there is current infection.

Control

When managing viral diseases in the individual animal, it is important for the clinician to look at the whole picture. Typically, suboptimal husbandry is a large contributing factor to the onset of disease; correction of husbandry deficits is one of the most important aspects of treatment. Diet, lack of an even and optimal thermal gradient so that the animal can properly thermoregulate, overcrowding, lack of appropriate hiding/perching areas, overhandling, inappropriate humidity, poor sanitary conditions, and other environmental stressors all play important roles in the onset of disease. Supportive care for the sick animal should be provided, including fluid therapy, treatment of bacterial or fungal secondary infections, and oxygen therapy if there is respiratory compromise. Reptiles have been shown to select higher temperatures in response to bacterial infection [7]; an ambulatory patient is able to select an appropriate temperature better than the clinician. Moribund animals may be unable to move to thermoregulate appropriately so the clinician must be even more attentive to thermal issues in these animals. Specific antiviral drugs, such as acyclovir for herpesviruses or cidofovir for adenoviruses, are discussed below; however, it is important to recognize that no pharmacokinetic, efficacy, or safety studies have been done for any of these drugs in any reptile species to date.

When dealing with a reptile collection, it is important to know the infection status for the pathogen of concern. When introducing new animals, they should be tested for diseases of concern, quarantined for a minimum of 90 days, and tested again before leaving quarantine. Quarantined animals should be cared for last by keepers or ideally cared for by separate keepers to prevent transmission of viruses by human vectors. Many viruses are host-adapted and cause mild to subclinical disease in their natural host species under optimal husbandry conditions; however, when an aberrant host species is infected, serious clinical disease may result. Therefore, mixed species enclosures should be avoided.

Selection of sources, if acquiring new animals, must be considered. When an animal is caught from the wild a large amount of inherent stress is involved. Typically, these animals are held in overcrowded, inappropriate conditions for what may be a long period of time until enough are collected to be shipped to a reptile importer. They are usually brought in without quarantine and a large amount of cross-contamination occurs with diverse reptile species from all over the world. A reptile importer, therefore, is an excellent source of fresh material for the reptile virologist, but is less than an ideal source for acquisition of animals for a collection. Captive bred animals, preferably from a breeder who works with a single species, are preferred. Another factor that plays a role in the development of disease is genetic diversity. Outbred, wild-type animals have a more robust immune system [8]. Typically, inbred color mutants are more susceptible to disease which makes them less desirable in a collection. Conversely, maintenance of a few color mutant animals as disease sentinels in a collection may be useful. In general, a closed collection with careful addition of new animals only for maintenance of genetic diversity is optimal.

If an animal is diagnosed with viral disease, the value of the animal must be weighed against the value of the collection and the transmissibility of the disease. Enveloped viruses, such as herpesviruses and paramyxoviruses, tend to be less stable in the environment. Prevention of transmission from one enclosure to another by human vectors is feasible with good hygiene practices, including not transferring food dishes or other cage furniture between enclosures, washing hands between enclosures, maintaining clean enclosures, using foot baths, and dealing with sick animals last. Conversely, nonenveloped viruses, such as adenoviruses and reoviruses, tend to be stable in the environment and present a greater challenge. Moving affected and exposed animals to a separate facility or culling should be considered.

Use of vaccines for prevention of viral infection in reptiles is an area that largely remains to be explored. An autogenous vaccine has been used for a poxvirus outbreak in Nile crocodiles (*Crocodylus niloticus*) with apparent efficacy [9]. A tortoise herpesvirus vaccine did not result in a significant increase in neutralizing antibody titers; virus challenge was not done [10]. This correlates well with mammals, in which poxvirus vaccines often are effective and herpesvirus vaccines are less effective. When considering vaccine design for an intracellular pathogen (eg, virus), it is important to elicit a cellular immune response. To activate CD8$^+$ T cells, the viral antigen needs to be presented by major histocompatibility complex (MHC) class I. MHC class I are expressed on the surface of most cells and present peptides that are found within the cell. Therefore, to elicit an optimal cellular immune response, a viral vaccine should be able to enter cells. Killed vaccines do not do this. Another option is the use of a modified live vaccine which has been passed through cell culture until it loses virulence. The main concern is the possibility of return to virulence, which has occurred with modified live vaccines [11]. Recombinant vaccines that use an unrelated avirulent virus—which expresses an immunogenic peptide from the pathogenic virus of concern—may hold the most promise for future vaccine development.

DNA viruses

The DNA viruses tend to have larger genomes. DNA viruses mutate slowly as compared with RNA viruses. Many DNA viruses have evolutionary histories that parallel those of their hosts [12,13]. This can provide interesting information about host as well as viral evolution and makes consensus PCR primer design easier. DNA viruses that have been identified in reptiles include herpesviruses, iridoviruses, adenoviruses, parvoviruses, poxviruses, and papillomaviruses (Table 1).

Herpesviruses

Herpesviruses are large, enveloped, double-stranded linear DNA viruses. These viruses are known for latency; if an animal is infected, it should be considered to be infected for life. Herpesviruses are divided into three subfamilies; all reptile herpesviruses that have been analyzed to date belong to the Alphaherpesvirinae subfamily [13–17]. Herpesviruses have been identified in chelonians, lizards, snakes, and recently, crocodilians.

Herpesviruses have been identified in many species of tortoises [18–26]. Susceptibility between tortoise species varies [18], but all tortoise species should be considered susceptible. Clinical signs can include the presence of diphtheric plaques in the oral cavity (Fig. 1), regurgitation, serous to mucopurulent nasal discharge, dyspnea, palpebral edema, ocular discharge, cervical edema, neurologic signs, anorexia, and lethargy. Intranuclear inclusions may be seen in infected cells (Fig. 2) and characteristic lesions include stomatitis, rhinitis, conjunctivitis, glossitis, pharyngitis, tracheitis, hepatitis, and pneumonia [18–21]. Differential diagnoses of upper respiratory infections include iridovirus, virus "X," and mycoplasmosis—a common bacterial upper respiratory tract disease of tortoises [27,28]. If stomatitis also is present, iridovirus and virus "X" would be higher on the differential list [29,30]. Diagnosis of tortoise herpesvirus has been made based on light microscopy, electron microscopy, virus isolation, PCR [14,31,32], immunohistochemistry [33], in situ hybridization [14], virus neutralization [34,35], and ELISA [34]. At least two serotypes exist that do not cross react [35] and must be taken into account when interpreting serology results.

Herpesvirus-like particles have been identified in freshwater turtles although the virus has not been isolated or further characterized. Herpesvirus-like particles were seen in tissues of two pacific pond turtles (*Clemmys marmorata*) [36], a painted turtle (*Chrysemys picta*) [37], a false map turtle (*Graptemys pseudogeographica*), and a Barbour's map turtle

Table 1
Clinical signs associated with viruses in reptiles

Signs	Viruses associated with signs
Stomatitis/pharyngitis/rhinitis in chelonians	Herpesvirus
	Iridovirus
	Picornavirus
	Reovirus
Tracheitis in chelonians	Herpesvirus
	Iridovirus
Hepatitis in chelonians	Herpesvirus
	Iridovirus
Enteritis in chelonians	Picornavirus
	Iridovirus
Pneumonia in chelonians	Picornavirus
Proliferative skin lesions in chelonians	Herpesvirus
	Papillomavirus
	Poxvirus
Hepatitis in snakes	Adenovirus
	Herpesvirus
Enteritis in snakes	Adenovirus
Neurologic signs in snakes	Paramyxovirus
	Reovirus
	Inclusion body disease
	Adenovirus
Pneumonia in snakes	Paramyxovirus
	Reovirus
Poor venom production in snakes	Herpesvirus
Neoplasia in snakes	Inclusion body disease
Oral proliferative lesions/neoplasia in lizards	Herpesvirus
Stomatitis in lizards	Herpesvirus
	Iridovirus
Enteritis in lizards	Adenovirus
Hepatitis in lizards	Herpesvirus
	Adenovirus
	Iridovirus
Pneumonia in lizards	Paramyxovirus
	Reovirus
	Iridovirus
Proliferative skin lesions in lizards	Poxvirus
	Herpesvirus
	Iridovirus
Hepatitis/enteritis in crocodilians	Adenovirus
	Flavivirus
Proliferative skin lesions in crocodilians	Poxvirus
Stomatitis in crocodilians	Flavivirus
Neurologic signs in crocodilians	Flavivirus

(*Graptemys barbouri*) [38]. Clinically, the turtles were anorectic and lethargic [36,38]. The painted turtle died following treatment for an aural abscess and did not exhibit any signs [37]. Histologically, hepatic and renal necrosis was seen in each case with eosinophilic intranuclear inclusion bodies in infected

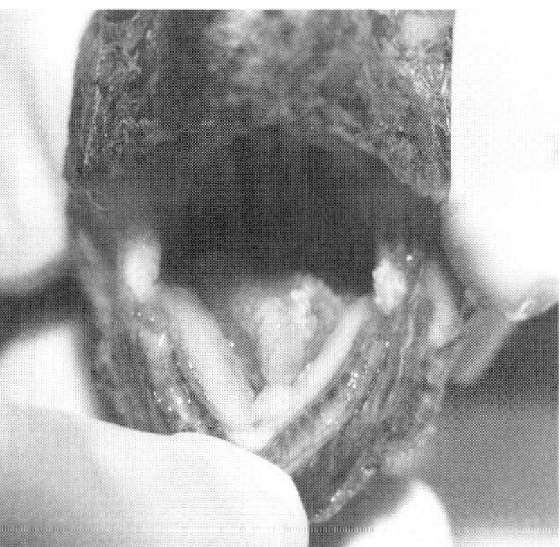

Fig. 1. Herpesvirus associated necrotizing stomatitis in a desert tortoise (*Gopherus agassizii*). (Courtesy of Roseanne Brown, DVM, El Cajon, California.)

cells. Recommended diagnostics include light and electron microscopy, virus isolation, PCR, and sequencing.

Gray patch disease; lung, eye, trachea disease (LETD); and fibropapillomatosis have been associated with herpesviruses in marine turtles. Gray patch disease causes circular papular skin lesions of juvenile green sea turtles (*Chelonia mydas*) [39]. LETD causes conjunctivitis, periglottal necrosis, tracheitis, and pneumonia in green sea turtles. Turtles show signs of respiratory disease and may have caseous exudate that covers the eyes [40]. Fibropapilloma-associated herpesviruses have been found in green turtles, loggerheads (*Caretta caretta*), and olive ridley turtles (*Lepidochelys olivacea*) [15,41]. Virus is found most often replicating within the papillomas and rarely is found elsewhere [15,42,43]. PCR has been used to identify DNA

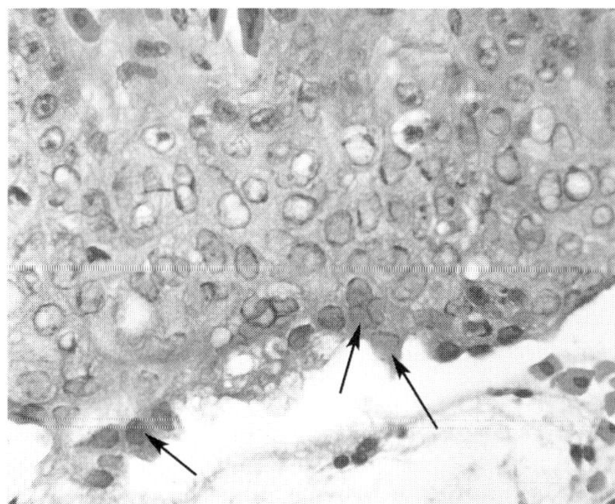

Fig. 2. Eosinophilic intranuclear inclusion bodies in glossal epithelial cells from the desert tortoise in Fig. 1. A few inclusions are indicated with arrows.

sequences and ELISA has been used to detect anti-LETD virus antibodies [44]. Diagnosis is made based on characteristic clinical lesions and visualization of intranuclear inclusions on light microscopy and virus particles on electron microscopy.

Herpesvirus has been reported in seven lizard species. Lacertid herpesvirus was identified by electron microscopy and associated with cutaneous papillomatous lesions in a European green lizard (*Lacertis viridis*) [45]. Iguanid herpesvirus was identified by electron microscopy in primary cell culture [46,47] and tissues [36] of iguanas (*Iguana iguana*). Experimental infections failed to produce a consistent pattern of lesions [46]. Agamid herpesvirus was detected by electron microscopy in tissues of two red-headed agamas (*Agama agama*) [48]. Herpesvirus was observed in, and sequenced from, a San Esteban chuckwalla (*Sauromalus varius*) that had hepatic necrosis [13], two Sudan plated lizards (*Gerrhosaurus major*), a black-lined plated lizard (*Gerrhosaurus nigrolineatus*), [16] and three green tree monitors (*Varanus prasinus*) that had proliferative stomatitis [17]. Virus was not isolated nor Koch's postulates fulfilled in the lacertid, agamid, chuckwalla, gerrhosaurid, or varanid herpesviruses; therefore, the relationship of the viruses to the lesions needs to be confirmed. Diagnosis can be made by light microscopy, electron microscopy, virus isolation, PCR, DNA sequencing, and in situ hybridization.

Herpesvirus-like particles were detected in the venom gland of Siamese cobras (*Naja siamensis*) [49] in association with poor quality venom and necrosis of glandular epithelial cells; they were associated with hepatic necrosis in two boa constrictors (*Boa constrictor*) [50]. Herpesvirus-like particles also were seen in the venom glands of Mojave rattlesnakes (*Crotalus scutulatus*) that had poor venom production (Elliott R. Jacobson, DVM, PhD, personal communication, 2003). The relationship of virus to lesions remains uncertain. Diagnosis was made based on light and electron microscopy. Virus isolation, PCR, and sequencing should be attempted to help confirm the diagnoses in future cases.

Herpesvirus recently was identified by PCR and sequencing in juvenile American alligators (*Alligator mississippiensis*) in association with lymphoid follicular inflammation of the cloaca [51]. Further studies are needed to determine if herpesvirus is the causative agent. Diagnosis can be made by biopsy, PCR, and sequencing.

PCR for reptile herpesviruses and serology testing for tortoise herpesvirus are being performed in the laboratories of Dr. Rachel Marschang at the Hohenheim University at Stuttgart in Germany, Dr. Elliott Jacobson at the University of Florida, and Dr. Silvia Blahak in Detmold, Germany. Acyclovir was effective against tortoise herpesvirus in vitro [52]; however, objective in vivo studies have not been performed. Although no pharmacokinetic studies have been performed in tortoises, one recommended dosage is 80 mg/kg/d orally [53], whereas other investigators suggested that 80 mg/kg, orally, every 8 hours, was more effective [54].

Topical 5% acyclovir ointment also has been used [19]. Although no efficacy studies have been performed, acyclovir also should be effective at reducing herpesvirus replication in other reptile species.

Iridoviruses

Iridoviruses are large, nonenveloped, double-stranded linear DNA viruses that may have the addition of a membrane envelope that is not necessary for infection. The family Iridoviridae consists of four genera. Chloriridovirus and Iridovirus typically infect insects, Lymphocystivirus infects fish, and Ranavirus is capable of infecting fish, amphibians, and reptiles [55]. The erythrocytic viruses make up another category of unclassified iridoviruses. These have not been sequenced but have been determined to be iridoviruses based on their morphologic and biochemical properties.

Ranavirus is becoming an important pathogen of chelonians. It has been reported in Hermann's tortoises (*Testudo hermanni*) [56–58], a Russian tortoise (*Agrionemys horsfieldi*) [55], box turtles (*Terrapene carolina*) [29,55], gopher tortoises (*Gopherus polyphemus*) [29,59], a soft-shelled turtle (*Trionyx sinensis*) [60], and a Burmese star tortoise (*Geochelone platynota*) [29,61]. Recently, there was an outbreak with mortality at several sites in the United States. The virus was isolated; sequencing of a portion of the major capsid protein gene demonstrated that it was related most closely to frog virus 3. An identical sequence was obtained from amphibians from two sites of infection; this suggested that amphibians may serve as a reservoir host of infection [61].

Clinical signs of infected turtles and tortoises are similar to those that are seen with herpesvirus or virus "X" [30,62], including necrotizing stomatitis, ocular and nasal discharge, conjunctivitis, and palpebral and cervical edema. "Red neck" or cervical edema with reddening was observed in the soft-shelled turtle [60]. Common histologic lesions with infection include necrotizing and ulcerative stomatitis, esophagitis, and tracheitis. Necrotizing conjunctivitis, splenitis, gastritis, enteritis, hepatitis, and vasculitis or thrombosis can occur. Necrosis of hematopoietic tissues is seen commonly in kidney, liver, and spleen and is characteristic of iridovirus infections. Basophilic intracytoplasmic inclusion bodies rarely can be observed within epithelial cells of the oral mucosa, esophagus, stomach, and trachea or within endothelial cells, macrophages, hepatocytes, and hematopoietic cells (Fig. 3).

Ranavirus also has been isolated from other reptiles, including a flat-tail gecko (*Uroplatus fimbriatus*) [63], a four-horned chameleon (*Chameleo quadricornis*) [63], and 10 green pythons (*Chondropython viridis*) [64]. Hepatitis with granulomatous lesions on the tongue and tail were observed in the gecko [63]; ulceration of the nasal mucosa and focal necrosis in the liver were observed in the pythons [64]. The relationship of virus to the etiology of the lesions is uncertain.

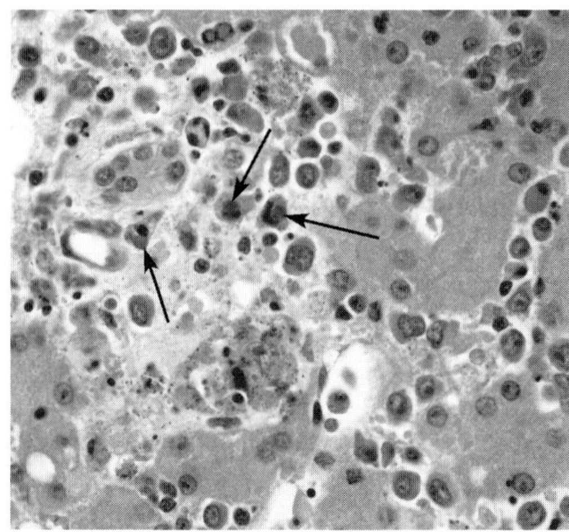

Fig. 3. Hematopoietic cell necrosis and basophilic intracytoplasmic inclusion bodies in the kidney of a gopher tortoise (*Gopherus polyphemus*) infected with iridovirus. Inclusion bodies are indicated with arrows.

Insect iridoviruses recently were isolated from reptiles which suggests they may be pathogenic to insectivorous reptiles. Iridovirus was isolated from two bearded dragons (*Pogona vitticeps*), a chameleon (*Chameleo quadricornis*), a frill necked lizard (*Chlamydosaurus kingii*) [65], and an iguana (*Iguana iguana*) [63]. No pathology was reported with these infections other than poxlike skin lesions on the frill necked lizard and a pneumonia in the bearded dragons. The relationship of virus to the etiology of these lesions is uncertain.

Snake, lizard and turtle erythrocytic viruses have been observed in many species since their first report in a gecko (*Tarentola mauritanica*) in 1914 [66]; however, little is known about these viruses. Variably shaped intracellular inclusions in erythrocytes and anemia often accompany the infection [67–69]. Early reports considered the organism to be protozoal and organisms were named *Toddia* or *Pirhemocyton*, depending on the presence of crystalline structures in the inclusions [70]. Many of the named species, however, were not confirmed with electron microscopy. Ultrastructural characterization later determined the inclusions to be the assembly site of viruses [71]. Experimental transmission studies by Alves de Matos et al [72] documented that the infection can resolve or become systemic and lead to death.

Diagnosis of ranavirus and iridovirus infections is made based on a combination of clinical signs, characteristic histologic lesions, PCR, virus isolation, and electron microscopy. Virus neutralization is available from the laboratory of Dr. Silvia Blahak in Detmold, Germany. Consensus primers have been described for PCR to detect a range of iridoviruses [73] and PCR is available in the laboratories of Dr. Rachel Marschang at the

Hohenheim University and Dr. Elliott Jacobson at the University of Florida. Treatment has not been determined in any species. Iridoviruses, like herpesviruses, have a viral thymidine kinase and, therefore, may be susceptible to acyclovir treatment. This has yet to be tested in vitro or in vivo. Diagnosis of erythrocytic virus infection is determined by the presence of intracytoplasmic inclusion bodies in erythrocytes on light microscopic examination followed by electron microscopy to visualize the virus particles.

Adenoviruses

Adenoviruses are medium-sized, nonenveloped, double-stranded linear DNA viruses. Taxonomically, all reptile adenoviruses belong to the genus *Atadenovirus* [74–77].

In reptiles, adenovirus-like particles have been identified in many hosts, including 10 snake species [78–86], 8 lizard species [77,87–90], Nile crocodiles (*Crocodylus niloticus*), [91], and a leopard tortoise (*Geochelone pardalis*) [92]. Lesions that are seen in reptiles in association with adenovirus-like agents include hepatitis [80,87,90,91], enteritis [78,79,88], esophagitis [81,89], splenitis [78], and encephalopathy [82]. The most commonly reported host species for adenoviral diseases are bearded dragons and kingsnakes. The clinical picture in an animal that has adenoviral disease often involves weight loss (Fig. 4), inappetence, and elevated liver enzymes. Histologically, adenoviruses often cause intranuclear inclusion bodies that distend the nucleus (Fig. 5). Few attempts have been made at antemortem diagnosis of adenoviral infection in reptiles; it is likely that increased surveillance would result in the recognition of additional disease syndromes and associated adenoviral types.

Methods that have been used for diagnosis of adenovirus infection in reptiles include virus isolation [86], electron microscopy [78], DNA in situ hybridization [84], plaque reduction neutralization [76], and a consensus sequence PCR [77]. Virus isolation has been successful from snakes [86] and a leopard tortoise [91], but multiple attempts at cultivation of the bearded dragon adenovirus have been unsuccessful (Elliott R. Jacobson, DVM, PhD, personal communication, 2003). Adenoviruses often are host specific and successful cultivation may depend on development of appropriate cell lines for cultivation. The diversity of reptile adenoviruses has not been well-explored and the degree of cross reactivity of neutralizing antibodies to reptile adenoviruses is not known. Virus neutralization is available from the laboratory of Dr. Rachel Marschang at Hohenheim University. Adenoviral consensus PCR and sequencing is available from the laboratories of Dr. Elliott Jacobson at the University of Florida and Dr. Rachel Marschang at Hohenheim University. The use of certain antiviral drugs (eg, cidofovir) has been shown to be effective for some adenoviral diseases [93]; this may prove to be the case with some reptile adenoviruses.

Fig. 4. Emaciated leopard gecko (*Eublepharis macularius*) infected with adenovirus. (Courtesy of Maud LaFortune, DMV, MSc, Lake Buena Vista, Florida.)

Parvoviruses

Parvoviruses are small, nonenveloped, single-stranded circular DNA viruses. The only reptile parvovirus that has been characterized is in the genus *Dependovirus*; they are dependent on coinfection with another viral species for replication [94]. Parvovirus-like particles have been seen on electron microscopy in the intestine or liver of four colubrid snake species [78,79,95], two boid snake species [94], and bearded dragons (*Pogona vitticeps*) [96]. These reptiles also had adenovirus-like particles. Although disease, including enteritis, hepatitis, neurologic signs, and pneumonia has been seen in parvovirus-infected lizards, it is difficult to attribute the disease to the parvoviral infection because the concurrent adenoviral coinfection was present. Typically, adenoviral inclusions are present in parvoviral infections in reptiles. No serologic or PCR diagnostic is available to the clinician for reptile parvoviruses.

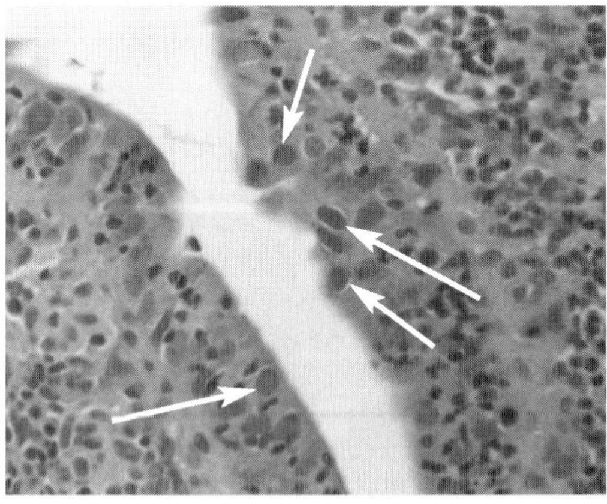

Fig. 5. Intranuclear inclusion bodies in epithelial cells of the small intestine. Inclusion bodies are indicated with arrows.

Poxviruses

Poxviruses are large, enveloped, double-stranded linear DNA viruses. No sequence information is available for any reptile poxvirus and none has been classified beyond the family Poxviridae.

Poxviruses have been identified in five species of crocodilians [9,97–99], a Hermann's tortoise (*Testudo hermanni*) [100], and two lizard species [101,102]. Most of these infections were associated with skin lesions that were scattered over the body surface and were particularly prominent on the head. Poxviral disease seems to be most common in Nile crocodiles. In a flap-necked chameleon (*Chameleo dilepis*), the poxviral inclusions were found in erythrocytes; skin lesions were not seen [102]. Poxviruses are associated with intracytoplasmic inclusion bodies. An autogenous vaccine has been used for a poxvirus outbreak in Nile crocodiles (*Crocodylus niloticus*) [9]. No reptile poxviruses have been isolated and no serologic or PCR diagnostics is available. The use of certain antiviral drugs (eg, cidofovir) has been shown to be effective for some poxviral diseases [103]; this may prove to be the case with some reptile poxviruses.

Papillomaviruses

Papillomaviruses are small, nonenveloped, double-stranded circular DNA viruses. No papillomavirus of reptiles has been characterized beyond the level of the family Papillomaviridae. Papillomavirus-like particles have been identified on electron microscopy in a green lizard (*Lacerta viridis*) [45], twist-necked turtles (*Platemys platycephala*) [104], and a Russian tortoise (*Agrionemys horsfieldii*) [105]. Papillomaviruses are not associated with inclusion bodies. The green lizard also had herpesvirus-like and reovirus-like particles and it is difficult to attribute clinical signs to the papilloma-like virus particles. In the twist-necked turtles, multi-focal, raised white oval lesions were seen on the head. These lesions regressed over time. Specific lesions were not reported to be associated with the virus in the Russian tortoise. No reptile papillomavirus has been isolated and no serologic or PCR diagnostics are available. The use of certain antiviral drugs (eg, topical cidofovir) has been shown to be effective for some papillomaviral lesions [106]; this may prove to be the case with some reptile papillomaviruses.

RNA viruses

The RNA viruses tend to have smaller genomes. RNA viruses mutate more rapidly than DNA viruses. Many RNA viruses mutate so rapidly that evolutionary change of the virus can be followed within a single host [107]. This makes consensus PCR primer design more difficult. RNA viruses that have been identified in reptiles include paramyxoviruses, reoviruses, retroviruses, flaviviruses, togaviruses, bunyaviruses, caliciviruses, rhabdoviruses, and picornaviruses.

Paramyxoviruses

Paramyxoviruses are medium-sized, enveloped, unsegmented, negative, single-stranded RNA viruses. All characterized reptile viruses in the family Paramyxoviridae are classified in the subfamily Paramyxovirinae; the genus *Ferlavirus* was proposed for characterized snake paramyxoviruses [108]. Studies of multiple isolates showed that the reptile paramyxoviruses, although related, tend to form two clusters [109,110]. The paramyxoviruses of reptiles all hemagglutinate. Reptile paramyxoviruses are fusogenic (form syncytia), much like reoviruses.

In reptiles, evidence of paramyxovirus infection has been found in eight lizard species [111–113] and several snake species [109,110,114]. Experimental infection of Aruba Island rattlesnakes (*Crotalus unicolor*) with an Aruba Island rattlesnake paramyxovirus caused fatal interstitial pneumonia and tracheitis. The virus was reisolated from the snakes; this fulfilled Koch's postulates [115]. Typically, a reptile that has paramyxoviral disease presents with pneumonia and neurologic signs; this is similar to the clinical picture that is seen with reovirus infection.

The histologic lesions of paramyxoviral disease resemble reoviral disease. Occasional intracytoplasmic inclusions may be seen in small numbers but are not always present [115]. Significant diversity is present among the reptile paramyxoviruses [109,110]. HI is the most common serologic test and is available from the laboratories of Dr. Elliott Jacobson at the University of Florida, Dr. Rachel Marschang at Hohenheim University, Dr. Silvia Blahak in Detmold, Germany, and the University of Tennessee. Reptile paramyxovirus PCR and sequencing is available from the laboratories of Dr. Elliott Jacobson at the University of Florida and Dr. Silvia Blahak in Detmold, Germany. The use of certain antiviral drugs (eg, ribavirin) has been shown to be effective for some paramyxoviral diseases in mammals [116]; this may prove to be the case with some reptile paramyxoviruses.

Reoviruses

Reoviruses are medium-sized, nonenveloped, segmented, double-stranded RNA viruses. All characterized reptile reoviruses are members of the genus *Orthoreovirus* [117,118]. Within this genus, reptile reoviruses seem to be related most closely to avian reoviruses and to fusogenic mammalian orthoreoviruses; this includes Nelson Bay reovirus from bats and a reovirus that causes encephalitis in baboons [117]. The reoviruses of reptiles do not hemagglutinate. Reptile reoviruses that have been cultured are fusogenic, much like paramyxoviruses.

In reptiles, evidence of reovirus infection has been found in 8 lizard species [45,111,112,119,120], and 11 snake species [118,119,121–125] (Jack M. Gaskin, DVM, PhD, personal communication, 2004), and 2 tortoise species [120,126]. Experimental infection of black ratsnakes (*Elaphe obsoleta*) with a Taiwan beauty snake (*Elaphe taenura*) and Moellendorff's ratsnake (*Elaphe moellendorffi*) reovirus caused fatal proliferative interstitial pneumonia and tracheitis. The virus was reisolated from the snakes, which fulfilled Koch's postulates [125]. A reptile that has reoviral disease typically presents with pneumonia and neurologic signs; this is similar to the clinical picture that is seen with paramyxovirus infection. In spur-thighed tortoises (*Testudo graeca*), epithelial necrosis of the tongue was associated with reoviral infection, as is seen in herpesviral, iridoviral, and picornaviral disease.

Diagnosis of reoviral disease is challenging because there are no inclusions and histologic lesions resemble paramyxoviral disease. There are several serologically distinct reptile reovirus isolates which indicates that significant diversity is present [119]. The degree of cross-reactivity of neutralizing antibodies to reptile reoviruses is not known. Because reptile reoviruses do not hemagglutinate, HI is not an option. Serologic diagnosis has been limited to virus neutralization [112], which is available from the laboratories of Dr. Rachel Marschang at Hohenheim University and Dr. Silvia Blahak in Detmold, Germany. Reoviral consensus PCR and sequencing is available from the laboratories of Dr. Elliott Jacobson at the University of Florida and Dr. Rachel Marschang at Hohenheim University.

Retroviruses

Retroviruses are small, enveloped RNA viruses that are reverse transcribed to DNA. They are divided into seven genera; the reptile retrovirus sequences cluster with the genera *Spumavirus* and *Gammaretrovirus* [127,128]. Retrovirus-like particles were observed first in reptiles by electron microscopy in a late passage cell culture of Asian pit viper (*Vipera russelli*) spleen cells. The viper had an edematous myxofibroma [129]. Subsequently, retrovirus-like particles were observed in various types of tumors of other reptiles, including a corn snake (*Elaphe guttata*) [130], two boa constrictors (*Boa constrictor*) [131], a four-lined chicken snake (*Elaphe obsoleta quadrivittata*) [132], green sea turtles (*Chelonia mydas*) [133], and four Burmese pythons (*Python molurus bivittatus*) [134]. No studies were performed to prove that retroviruses were the causative agent of the tumors, although retroviruses in other species are known to be oncogenic.

Retrovirus particles also have been identified in healthy reptiles, including the venom glands of *Bothrops jururucussu* [135], green sea turtles [133], the tuatara (*sphenodon punctatus*) [136], Komodo dragon (*Varanus komodoensis*) [137], the South American pit viper (*Bothrops jararaca*) [128], a red-eared slider turtle (*Chrysemys scripta*) [128], and six crocodilians from three families [137].

Retrovirus-like particles have been documented in snakes that were infected with inclusion body disease (IBD). IBD is a disease that affects snakes in the family Boidae and typically causes neurologic disease that is characterized by stargazing, tremors, disequilibrium, disorientation, and paresis. Chronic regurgitation can occur and snakes often are more susceptible to secondary infections and neoplasia [138]. C-type retrovirus-like particles were observed—on electron microscopy—budding from the endoplasmic reticulum and plasma membrane in infected snakes; a retrovirus was believed to be the causative agent [138]. Experimental transmission studies that infected boa constrictors with infected liver homogenate demonstrated the presence of retrovirus-like particles after infection that were not seen on preinfection biopsy samples [139]. Attempts to identify a retrovirus that is responsible for IBD has led to identification of new retroviruses, but none have proven to be causative of IBD [140]. IBD is diagnosed histologically by the detection of large eosinophilic intracytoplasmic inclusions in biopsies of most major organs. More work needs to be done to determine the etiology of IBD before serologic or PCR testing will be available.

Flaviviruses

Flaviviruses are small, enveloped, unsegmented, positive-sense, single-stranded RNA viruses that typically are arthropod borne. As a group, flaviviruses include several serious diseases in humans, including West Nile virus, Japanese encephalitis, hepatitis C, yellow fever, and dengue fever. All members of the family Flaviviridae, for which there is evidence of infection in reptiles, are in the genus *Flavivirus*.

West Nile virus (WNV) has been associated with encephalitis, stomatitis, hepatitis, and high death losses in American alligators (*Alligator mississippiensis*) [141,142]. All alligators that developed clinical disease in the first outbreak were kept at warmer than normal temperatures in dark, igloolike enclosures. There were high levels of mortality in some enclosures and none in others. Water was changed too frequently to allow for successful breeding of mosquitoes which suggested alligator–alligator transmission within enclosures. Clinical WNV disease has not been identified in wild alligators. A crocodile monitor (*Varanus salvadorii*) with neurologic signs also was infected with WNV [143]. Serologic evidence of infection in the absence of apparent clinical disease was found in Nile crocodiles (*Crocodylus niloticus*) [144] and a Caspian pond turtle (*Mauremys* [*Clemmys*] *caspica*) [145]. Small, but detectable, viral loads were seen after infection of green iguanas (*Iguana iguana*), Florida garter snakes (*Thamnophis sirtalis sirtalis*), and red-eared sliders (*Trachemys scripta*) with WNV [146]. PCR for WNV is available in several laboratories.

A flavivirus was isolated from a leopard tortoise (*Geochelone pardalis*) in association with fatal hemolysis and wasting [147].

Japanese grass lizards (*Takydromus tachydromoides*) and Japanese five lined skinks (*Eumeces latiscutatus*) were viremic posthibernation when experimentally infected with Japanese encephalitis virus prehibernation [148]. No sign of disease was noted in these lizards. Several reptiles were found to be seropositive for Japanese encephalitis virus [149].

Green lizards (*Lacerta viridis*) and sand lizards (*Lacerta agilis*) were infected experimentally with tick borne encephalitis virus [150,151]. No sign of disease was noted.

Togaviruses

Togaviruses are small, enveloped, unsegmented, positive-sense, single-stranded RNA viruses that typically are arthropod borne. As a group, togaviruses include several serious diseases in humans, including eastern equine encephalitis (EEE), western equine encephalitis (WEE), Venezuelan equine encephalitis, and rubella. All members of the family Togaviridae for which there is evidence of infection in reptiles belong to the genus *Alphavirus*.

Experimental infection of five turtle species and four snake species with EEE resulted in viremia for up to 28 days [152]. No sign of disease was seen in these animals. There are multiple reports of EEE seropositivity in reptile species [149].

Texas tortoises (*Gopherus berlandieri*) that were infected experimentally with WEE showed prolonged viremia and a limited humoral immune response [153]. Although no specific lesions were associated with infection, there was a high mortality rate in the group. Garter snakes (*Thamnophis* spp) also have been infected experimentally with WEE [154]. There are several reports of WEE seropositivity in reptiles [149]. Overall, zoonotic togaviruses have not been found to be a major cause of disease in reptiles.

Bunyaviruses

Bunyaviruses are medium-sized, enveloped, segmented, negative-sense, single-stranded RNA viruses that typically are arthropod borne. As a family, bunyaviruses include several serious diseases in humans, including Rift Valley fever, Crimean-Congo hemorrhagic fever, and Four Corners disease.

A bunyavirus was isolated from a spiny softshell turtle (*Apalone* [*Trionyx*] *spinifer*) [155]. This agent was capable of killing suckling mice. Serologically, the virus seemed to be related most closely to Cache Valley virus and Tenshaw virus serotypes (Family Bunyaviridae; genus *Bunyavirus*; species *bunyamweravirus*). Cache Valley virus causes stillbirths and congenital malformations in sheep.

Caliciviruses

Caliciviruses are small, nonenveloped, positive-sense, single-stranded linear RNA viruses. Caliciviruses have been isolated from Aruba island rattlesnakes (*Crotalus unicolor*), a rock rattlesnake (*Crotalus lepidus*), and an eyelash viper (*Bothriechis* [*Bothrops*] *schlegeli*) [156]. The caliciviruses that have been identified from reptiles belong to the genus *Vesivirus*. No specific disease syndrome was associated with the virus in the snakes from which the viruses were isolated and no disease was seen in experimentally infected snakes or pigs. Typically, caliciviruses do not cause inclusion bodies. Calicivirus isolates from marine mammals were serologically indistinguishable from these reptilian isolates [157]. A PCR protocol has been designed that will amplify a reptile calicivirus [158].

Rhabdoviruses

Rhabdoviruses are medium-sized, enveloped, negative-sense single-stranded linear RNA viruses. As a group, rhabdoviruses cause several diseases of human and veterinary concern, including rabies and vesicular stomatitis. The isolated rhabdoviruses of reptiles have not been classified beyond the level of the family (Rhabdoviridae). Vesicular stomatitis virus, for which there is only serologic evidence in reptiles, is in the genus Vesiculovirus. Three rhabdoviruses have been isolated from reptiles— Marco, Chaco, and Timbo viruses [159]. All were isolated from a South American teiid lizard, *Amieva amieva*. There also has been one isolation of Chaco virus from another South American teiid, *Kentropyx calcarata*. An uncharacterized rhabdovirus has been seen on electron microscopy of a third South American teiid, the caiman lizard (*Dracaena guianesis*) (Allan Pessier, DVM, personal communication, 2004). These viruses do not seem to cause clinically significant disease in the host species. Serologic evidence for vesicular stomatitis virus infection has been found in red-bellied water snakes (*Natrix erythrogaster*) [155] and a spiny softshell turtle (*Apalone* [*Trionyx*] *spinifer*) [160]. Rhabdoviruses may be associated with intracytoplasmic inclusions. No serologic or PCR diagnostic is available.

Picornaviruses

Picornaviruses are small, nonenveloped, positive-sense, single-stranded linear RNA viruses. Picornavirus-like particles have been seen on electron microscopy in the duodenum and spleen of a boa constrictor and the duodenum of an Aesculapian snake (*Elaphe longissima*); both had necrotic enteritis [78]. Both snakes also had adenovirus-like particles. The possibility exists that these were adenovirus-dependent parvoviruses, which are similarly-sized nonenveloped viruses. A small, nonenveloped, single-stranded linear RNA virus also has been isolated from tortoises that had necrotizing stomatitis, pharyngitis, conjunctivitis, rhinitis, pneumonia, enteritis, and ascites, and has been called "virus X" [30]. Typically, picornaviruses do not result in inclusion bodies. This virus is most consistent with, and is believed to be, a picornavirus [126]. The virus has been cultured successfully and virus neutralization testing is available in the laboratory of Dr. Rachel Marschang at Hohenheim University.

References

[1] Lock BA, Green LG, Jacobson ER, et al. Use of an ELISA for detection of antibody responses in Argentine boa constrictors (*Boa constrictor occidentalis*). Am J Vet Res 2003; 64:388–95.

[2] Zapata AG, Varas A, Torroba M. Seasonal variations in the immune system of lower vertebrates. Immunol Today 1992;13:142–7.

[3] el Masri M, Saad AH, Mansour MH, et al. Seasonal distribution and hormonal modulation of reptilian T cells. Immunobiology 1995;193:15–41.

[4] Munoz FJ, de la Fuente M. The immune response of thymic cells from the turtle *Mauremys caspica*. J Comp Physiol [B] 2001;171:195–200.

[5] Saad AH, el Deeb S. Immunological changes during pregnancy in the viviparous lizard, *Chalcides ocellatus*. Vet Immunol Immunopathol 1990;25:279–86.

[6] Mondal S, Rai U. Dose and time-related in vitro effects of glucocorticoid on phagocytosis and nitrite release by splenic macrophages of wall lizard *Hemidactylus flaviviridis*. Comp Biochem Physiol Toxicol Pharmacol 2002;132:461–70.

[7] Muchlinski AE, Gramajo R, Garcia C. Pre-existing bacterial infections, not stress fever, influenced previous studies which labeled *Gerrhosaurus major* an afebrile lizard species. Comp Biochem Physiol A 1999;124:353–7.

[8] McClelland EE, Penn DJ, Potts WK. Major histocompatibility complex heterozygote superiority during coinfection. Infect Immun 2003;71:2079–86.

[9] Horner RF. Poxvirus in farmed Nile crocodiles. Vet Rec 1988;122:459–62.

[10] Marschang RE, Milde K, Bellavista M. Virus isolation and vaccination of Mediterranean tortoises against a chelonid herpesvirus in a chronically infected population in Italy. Dtsch Tierarztl Wochenschr 2001;108:376–9.

[11] Opriessnig T, Halbur PG, Yoon KJ, et al. Comparison of molecular and biological characteristics of a modified live porcine reproductive and respiratory syndrome virus (PRRSV) vaccine (ingelvac PRRS MLV), the parent strain of the vaccine (ATCC VR2332), ATCC VR2385, and two recent field isolates of PRRSV. J Virol 2002;76: 11837–44.

[12] McGeoch DJ, Davison AJ. The molecular evolutionary history of the herpesviruses. In: Domingo E, Webster R, Holland J, editors. Origin and evolution of viruses. San Diego (CA): Academic Press; 1999. p. 441–65.

[13] Wellehan JFX, Jarchow JL, Reggiardo C, et al. A novel herpesvirus associated with hepatic necrosis in a San Esteban chuckwalla, *Sauromalus varius*. J Herpetol Med Surg 2003;13(3): 15–9.

[14] Teifke JP, Löhr CV, Marschang RE, et al. Detection of chelonid herpesvirus DNA by nonradioactive in situ hybridization in tissues from tortoises suffering from stomatitis-rhinitis complex in Europe and North America. Vet Pathol 2000;37:377–85.

[15] Quackenbush SL, Work TM, Balazs GH, et al. Three closely related herpesviruses are associated with fibropapillomatosis in marine turtles. Virology 1998;246(2):392–9.

[16] Wellehan JFX, Nichols DK, Li LL, et al. Three novel herpesviruses associated with stomatitis in Sudan plated lizards (*Gerrhosaurus major*) and a black-lined plated lizard (*Gerrhosaurus nigrolineatus*). J Zoo Wildl Med 2004;35(1):50–4.

[17] Wellehan J, Johnson AJ, Jacobson E, et al. Novel herpesviruses associated with stomatitis in lizards. In: Proceedings of the Association of Reptilian and Amphibian Veterinarians. Minneapolis: 2003.

[18] Jacobson ER, Clubb S, Gaskin JM, et al. Herpesvirus-like infection in Argentine tortoises. J Am Vet Med Assoc 1985;187:1227–9.

[19] Cooper JE, Gschmeissner S, Bone RD. Herpes-like virus particles in necrotic stomatitis of tortoises. Vet Rec 1988;123:554.

[20] Heldstab A, Bestetti G. Herpesviridae causing glossitis and meningoencephalitis in land tortoises (Testudo hermanii). Herpetopathologia 1989;1(2):5–9.

[21] Drury SEN, Gough RE, McArthur S, et al. Detection of herpes-like and papilloma-like particles associated with diseases of tortoises. Vet Rec 1998;143:639.

[22] Muro J, Ramis A, Pastor J, et al. Chronic rhinitis associated with herpesviral infection in captive spur-thighed tortoises from Spain. J Wildl Dis 1998;34(3):487–95.

[23] Une Y, Uemura K, Nakano Y, et al. Herpesvirus infection in tortoises (*Malacochersus tornieri* and *Testudo horsfieldii*). Vet Pathol 1999;36:624–7.

[24] Pettan-Brewer KCB, Drew ML, Ramsay E, et al. Herpesvirus particles associated with oral and respiratory lesions in a California desert tortoise (*Gopherus agassizii*). J Wildl Dis 1996; 32:521–6.

[25] Harper PAW, Hammond DC, Hewschele W. A herpesvirus-like agent associated with a pharyngeal abscess in a desert tortoise. J Wildl Dis 1982;18:491–4.

[26] Martínez-Silvestre A, Majó N, Ramis A. Clinical case: Herpesviral disease in a desert tortoise (*Gopherus agassizii*) [Caso clínico: herpesvirosis en tortuga de desierto Americana (*Gopherus agassizii*)]. Clínica Veterinaria de Pequn˜os Animales 1999;19(2):99–106 [in Spanish].

[27] Brown MB, Schumacher IM, Klein PA, et al. *Mycoplasma agassizii* causes upper respiratory tract disease in the desert tortoise. Infect Immun 1994;62:4580–6.

[28] Mathes KA, Jacobson ER, Blahak S, et al. Mycoplasma and herpes virus detection in European terrestrial tortoises in France and Morocco. In: Proceedings of the Association of Reptilian and Amphibian Veterinarians. Orlando: 2001. p. 97–9.

[29] Johnson AJ, Jacobson ER. Iridovirus infections of turtles and tortoises. In: Proceedings of the 29th Annual Symposium of the Desert Tortoise Council. Las Vegas: 2004. p. 24–5.

[30] Marschang RE, Ruemenapf TH. Virus "X" characterizing a new viral pathogen in tortoises. In: Proceedings of the Association of Reptilian and Amphibian Veterinarians. Reno: 2002, p. 101–2.

[31] Une Y, Murakami M, Uemura K, etc. Polymerase chain reaction (PCR) for the detection of herpesvirus in tortoises. J Vet Med Sci 2000;62(8):905–7.

[32] Origgi FC, Romero CH, Bloom DC, et al. Experimental transmission of a herpesvirus in greek tortoises (*Testudo graeca*). Vet Pathol 2004;41:50–61.

[33] Origgi FC, Klein PA, Tucker SJ, et al. Application of immunoperoxidase-based techniques to detect herpesvirus infection in tortoises. J Vet Diagn Invest 2003;15:133–40.

[34] Origgi FC, Klein PA, Mathes K, et al. Enzyme-linked immunosorbent assay for detecting herpesvirus exposure in Mediterranean tortoises (spur-thighed tortoise [*Testudo graeca*] and Hermann's tortoise [*Testudo hermanni*]). J Clin Micro 2001;39:3156–63.

[35] Marschang RE, Frost JW, Gravendyck M, et al. Comparison of 16 chelonid herpesviruses by virus neutralization tests and restriction endonuclease digestion of viral DNA. J Vet Med B 2001;48:393–9.

[36] Frye FL, Oshiro LS, Dutra FR, et al. Herpesvirus-like infection in two Pacific pond turtles. J Am Vet Med Assoc 1977;171(9):882–3.

[37] Cox WR, Rapley WA, Barker LK. Herpesvirus-like infection in a painted turtle (*Chrysemys picta*). J Wildl Dis 1980;16(3):445–9.

[38] Jacobson ER, Gaskin JM, Wahlquist H. Herpesvirus-like infection in map turtles. J Am Vet Med Assoc 1982;181(11):1322–4.

[39] Rebell H, Rywlin A, Haines HA. Herpesvirus-type agent associated with skin lesions of green turtles in aquaculture. Am J Vet Res 1975;36:1221–4.

[40] Jacobson ER, Gaskin JM, Roelke M, et al. Conjunctivitis, tracheitis, and pneumonia associated with herpesvirus infection of green sea turtles. J Am Vet Med Assoc 1986;189: 1020–3.

[41] Herbst LH, Jacobson ER, Klein PA, et al. Comparative pathology and pathogenesis of spontaneous and experimentally induced fibropapillomas of green turtles (*Chelonia mydas*). Vet Pathol 1999;36(6):551–64.

[42] Lu Y, Wang Y, Aguirre AA, et al. RT-PCR detection of the expression of the polymerase gene of a novel reptilian herpesvirus in tumor tissues of green turtles with fibropapilloma. Arch Virol 2003;148(6):1155–63.

[43] Lackovitch JK, Brown DR, Homer BL, et al. Association of herpesvirus with fibropapillomatosis of the green turtle *Chelonia mydas* and the loggerhead turtle *Caretta caretta* in Florida. Dis Aquat Organ 1999;37:89–97.

[44] Coberly SS, Condit RC, Herbst LH, et al. Identification and expression of immunogenic proteins of a disease-associated marine turtle herpesvirus. J Virol 2002;76(20):10553–8.

[45] Raynaud A, Adrian M. Cutaneous lesions with papillomatous structure associated with viruses in the green lizard (*Lacerta viridis* Laur.). C R Acad Sci Hebd Seances Acad Sci D 1976;283:845–7.

[46] Clark HF, Karzon DT. Iguana virus, a herpes-like virus isolated from cultured cells of a lizard, *Iguana iguana*. Infect Immun 1972;5:559–69.

[47] Zeigel RF, Clark HF. Electron microscopy observations on a new herpes-type virus isolated from *Iguana iguana* and propagated in reptilian cells in vitro. Infect Immun 1972; 5(4):570–82.

[48] Watson GL. Herpesvirus in red-headed (common) agamas (*Agama agama*). J Vet Diagn Invest 1993;5:444–5.

[49] Simpson CF, Jacobson ER, Gaskin JM. Herpesvirus-like infection of the venom gland of Siamese cobras. J Am Vet Med Assoc 1979;175(9):941–3.

[50] Hauser B, Mettler F, Rubel A. Herpesvirus-like infection in two young boas. J Comp Pathol 1983;93(4):515–9.

[51] Govett PD, Harms CA, Johnson A, et al. Lymphoid follicular cloacal inflammation in juvenile alligators (*Alligator mississippiensis*). In: Proceedings of the International Association for Aquatic Animal Medicine. Galveston, Texas: 2004. p. 153–4.

[52] Marschang RE, Gravendyck M, Kaleta EF. Herpesviruses in tortoises: investigations into virus isolation and the treatment of viral stomatitis in *Testudo hermanni* and *T. graeca*. J Vet Med B 1997;44:385–94.

[53] Schumacher J. Viral diseases. In: Mader DR, editor. Reptile medicine and surgery. Philadelphia: WB Saunders; 1996. p. 224–34.

[54] Wilkinson R. Therapeutics. In: McArthur S, Wilkinson R, Meyer J, editors. Medicine and surgery of tortoises and turtles. Ames (IA): Blackwell Publishing; 2004. p. 465–85.

[55] Mao J, Hedrick RP, Chinchar VG. Molecular characterization, sequence analysis and taxonomic position of newly isolated fish iridoviruses. Virology 1997;229:212–20.

[56] Heldstab A, Bestetti G. Spontaneous viral hepatitis in a spur-tailed Mediterranean land tortoise (*Testudo hermanni*). J Zoo Anim Med 1982;13:113–20.

[57] Muller M, Zangger N, Denzler T. An iridovirus outbreak in Hermann's tortoises (*Testudo hermanni* hermanni). [Iridovirus-epidemie bei der griechischen Landschildkrote (*Testudo hermanni hermanni*)]. Verhandl Ber 30. Int Symp Erkr Zoo-und Wildtiere, Sofia: 1988. p. 271–4 [in German].

[58] Marschang RE, Becher P, Posthaus H, et al. Isolation and characterization of an iridovirus from Hermann's tortoises (*Testudo hermanni*). Arch Virol 1999;144:1909–22.

[59] Westhouse RA, Jacobson ER, Harris RK, et al. Respiratory and pharyngo-esophageal iridovirus infection in a gopher tortoise (*Gopherus polyphemus*). J Wildl Dis 1996;32:682–6.

[60] Chen Z, Zheng J, Jian Y. A new iridovirus isolated from soft-shelled turtle. Virus Res 1999; 63:147–51.

[61] Johnson AJ, Norton TM, Wellehan JFX, et al. Iridovirus outbreak in captive Burmese star tortoises (*Geochelone platynota*). In: Proceedings of the Association of Reptilian and Amphibian Veterinarians. Naples: 2004. p. 143–4.

[62] Origgi FC, Romero CH, Bloom DC, et al. Experimental transmission of a herpesvirus in Greek tortoises (*Testudo graeca*). Vet Pathol 2004;41(1):50–61.

[63] Marschang RE, Becher P, Braun S. Isolation of iridoviruses from three different lizard species. In: Proceedings of the Association of Reptilian and Amphibian Veterinarians. Reno: 2002. p. 99–100.

[64] Hyatt AD, Williamson M, Coupar BEH, et al. First identification of a *Ranavirus* from green pythons (*Chondropython viridis*). J Wildl Dis 2002;38(2):239–52.

[65] Just F, Essbauer S, Ahne W, et al. Occurrence of an invertebrate iridescent-like virus (*Iridoviridae*) in reptiles. J Vet Med B 2001;48:685–94.

[66] Chatton E, Blanc G. On a novel hemoparasite *Pirhemocyton tarentolae* of the gecko *Tarentola mauritanica*, and the systemic changes that it causes. (Sur un hematozoaire nouveau, *Pirhemocyton tarentolae*, du gecko *Tarentola mauritanica*, et sur les alteracions globulaires quil determine.) Compt Rend Soc Biol 1914;77:496–8 [in French].

[67] Marquardt WC, Yaeger RG. The structure and taxonomic status of *Toddia* from the cottonmouth snake *Agkistrodon piscivorous leucostoma*. J Protozool 1967;14:726–31.

[68] Smith TG, Desser SS, Hong H. Morphology, ultrastructure and taxonomic status of *Toddia* sp. in northern water snakes (*Nerodia sipedon sipedon*) from Ontario, Canada. J Wildl Dis 1994;30:169–75.

[69] Johnsrude JD, Raskin RE, Hoge AY, et al. Intraerythrocytic inclusions associated with iridoviral infection in a fer de lance (*Bothrops moojeni*) snake. Vet Pathol 1997;34:235–8.

[70] Telford SR. Haemoparasites of reptiles. In: Hoff GL, Frye FL, Jacobson ER, editors. Diseases of amphibians and reptiles. New York: Plenum Publishing Co; 1984. p. 482–7.

[71] Stehbens WE, Johnston MRL. The viral nature of *Pirhemocyton tarentolae*. J Ultrastruct Res 1966;15:543–54.

[72] Alves de Matos AP, Paperna I, Crespo E. Experimental infection of lacertids with lizard erythrocytic viruses. Intervirology 2002;45:150–9.

[73] Gould AR, Hyatt AD, Hengstberger SH, et al. A polymerase chain reaction (PCR) to detect epizootic haematopoietic necrosis virus and Bohle iridovirus. Dis Aquat Org 1995;22: 211–5.

[74] Benkö M, Élö P, Ursu K, et al. First molecular evidence for the existence of distinct fish and snake adenoviruses. J Virol 2002;76:10056–9.

[75] Farkas SL, Benkö M, Élö P, et al. Genomic and phylogenetic analyses of an adenovirus isolated from a corn snake (*Elaphe guttata*) imply common origin with the members of the proposed new genus *Atadenovirus*. J Gen Virol 2002;83:2403–10.

[76] Marschang RE, Michling R, Benkö M, et al. Evidence for wide-spread atadenovirus infection among snakes. In: Proceedings of the 6th International Congress of Veterinary Virology. ZOOPOLE développement – ISPAIA, Ploufragan, France. 2003. p. 152.

[77] Wellehan JFX, Johnson AJ, Harrach B, et al. Detection and analysis of six lizard adenoviruses by consensus primer PCR provides further evidence of a reptilian origin for atadenovirus. Journal of Virology 2004;78, in press.

[78] Heldstab A, Bestetti G. Virus associated gastrointestinal diseases in snakes. J Zoo Anim Med 1984;14:118–28.

[79] Wozniak EJ, DeNardo DF, Brewer A, et al. Identification of adenovirus- and dependovirus-like agents in an outbreak of fatal gastroenteritis in captive born California mountain kingsnakes, *Lampropeltis zonata multicincta*. J Herpetol Med Surg 2000;10(2): 4–7.

[80] Schumacher J, Jacobson ER, Burns R, et al. Adenovirus-like infection in two rosy boas (*Lichanura trivirgata*). J Zoo Wildl Med 1994;25:461–5.

[81] Raymond JT, Garner MM, Murray S, et al. Oroesophageal adenovirus-like infection in a palm viper, *Bothriechis marchi*, with inclusion body-like disease. J Herpetol Med Surg 2002;12(3):30–2.

[82] Raymond JT, Lamm M, Nordhausen R, et al. Degenerative encephalopathy in a costal mountain kingsnake (*Lampropeltis zonata multifasciata*) due to adenoviral-like infection. J Wildl Dis 2003;39:431–6.

[83] Juhasz A, Ahne W. Physicochemical properties and cytopathogenicity of an adenovirus-like agent isolated from a corn snake (*Elaphe guttata*). Arch Virol 1992;130:429–39.

[84] Perkins LEL, Campagnoli RP, Harmon BG, et al. Detection and confirmation of reptilian adenovirus infection by *in situ* hybridization. J Vet Diagn Invest 2001;13:365–8.

[85] Ogawa M, Ahne W, Essbauer S. Reptilian viruses: adenovirus-like agent isolated from royal python (*Python regius*). J Vet Med B 1992;39:732–6.

[86] Jacobson ER, Gaskin JM, Gardiner CH. Adenovirus-like infection in a boa constrictor. J Am Vet Med Assoc 1985;187:1226–7.

[87] Jacobson ER, Kollias GV. Adenovirus-like infection in a savannah monitor. Journal of Zoo Animal Medicine 1986;17:149–51.

[88] Kinsel MJ, Barbiers RB, Manharth A, et al. Small intestinal adeno-like virus in a mountain chameleon (*Chameleo montium*). J Zoo Wildl Med 1997;28:498–500.

[89] Jacobson ER, Gardiner CH. Adeno-like virus in esophageal and tracheal mucosa of a Jackson's chameleon (*Chamaeleo jacksoni*). Vet Pathol 1990;27:210–2.

[90] Julian AF, Durham PJK. Adenoviral hepatitis in a female bearded dragon (*Amphibolurus barbatus*). N Z Vet J 1982;30:59–60.

[91] Jacobson ER, Gardiner CH, Foggin CM. Adenovirus-like infection in two Nile crocodiles. J Am Vet Med Assoc 1984;185:1421–2.

[92] Wilkinson R. Clinical pathology. In: McArthur S, Wilkinson R, Meyer J, editors. Medicine and surgery of tortoises and turtles. Oxford (UK): Blackwell Press; 2004. p. 141–86.

[93] Romanowski EG, Yates KA, Gordon YJ. Antiviral prophylaxis with twice daily topical cidofovir protects against challenge in the adenovirus type 5/New Zealand rabbit ocular model. Antiviral Res 2001;52:275–80.

[94] Farkas SL, Zadori Z, Benko M, et al. A parvovirus isolated from a royal python (*Python regius*) is a member of the genus dependovirus. J Gen Virol 2004;85:555–61.

[95] Ahne W, Scheinert P. Reptilian viruses: isolation of parvovirus-like particles from corn snake *Elaphe guttata (Colubridae)*. Zentralbl Veterinarmed B 1989;36:409–12.

[96] Jacobson ER, Kopit W, Kennedy FA, et al. Coinfection of a bearded dragon (*Pogona vitticeps*) with adenovirus- and dependovirus-like particles. Vet Pathol 1996;33: 429–39.

[97] Jacobson ER, Popp JA, Shields RP, et al. Poxlike skin lesions in captive caimans. J Am Vet Med Assoc 1979;175:937–40.

[98] Buenviaje GN, Ladds PW, Melville L. Poxvirus infection in two crocodiles. Aust Vet J 1992;69:15–6.

[99] Ramos MC, Coutinho SD, Matushima ER, et al. Poxvirus dermatitis outbreak in farmed Brazilian caimans (*Caiman crocodilus yacare*). Aust Vet J 2002;80:371–2.

[100] Oros J, Rodriguez JL, Deniz S, et al. Cutaneous poxvirus-like infection in a captive Hermann's tortoise (*Testudo hermanni*). Vet Rec 1998;143:508–9.

[101] Stauber E, Gogolewski R. Poxvirus dermatitis in a tegu lizard (*Tubinambis teguixin*). J Zoo Wildl Med 1990;21:228–30.

[102] Jacobson ER, Telford SR. Chlamydial and poxvirus infections of circulating monocytes of a flap-necked chameleon (*Chamaeleo dilepis*). J Wildl Dis 1990;26:572–7.

[103] Smee DF, Bailey KW, Wong MH, et al. Effects of cidofovir on the pathogenesis of a lethal vaccinia virus respiratory infection in mice. Antiviral Res 2001;52(1):55–62.

[104] Jacobson ER, Gaskin JM, Clubb S, et al. Papilloma-like virus infection in Bolivian side-neck turtles. J Am Vet Med Assoc 1982;181:1325–8.

[105] Drury SE, Gough RE, McArthur S, et al. Detection of herpesvirus-like and papillomavirus-like particles associated with diseases of tortoises. Vet Rec 1998;143(23):639.

[106] Matteelli A, Beltrame A, Graifemberghi S, et al. Efficacy and tolerability of topical 1% cidofovir cream for the treatment of external anogenital warts in HIV-infected persons. Sex Transm Dis 2001;28(6):343–6.

[107] Moya A, Holmes EC, Gonzalez-Candelas F. The population genetics and evolutionary epidemiology of RNA viruses. Nat Rev Microbiol 2004;2:279–88.

[108] Kurath G, Batts WN, Ahne W, et al. Complete genome sequence of Fer-de-Lance virus reveals a novel gene in reptilian paramyxoviruses. J Virol 2004;78:2045–56.

[109] Ahne W, Batts WN, Kurath G, et al. Comparative sequence analyses of sixteen reptilian paramyxoviruses. Virus Res 1999;63:65–74.

[110] Franke J, Essbauer S, Ahne W, et al. Identification and molecular characterization of 18 paramyxoviruses isolated from snakes. Virus Res 2001;80:67–74.

[111] Gravendyck M, Ammermann P, Marschang RE, et al. Paramyxoviral and reoviral infections of iguanas on Honduran Islands. J Wildl Dis 1998;34:33–8.

[112] Marschang RE, Donahoe S, Manvell R, et al. Paramyxovirus and reovirus infections in wild-caught Mexican lizards (*Xenosaurus* and *Abronia* spp.). J Zoo Wildl Med 2002;33(4): 317–21.

[113] Jacobson ER, Origgi F, Pessier AP, et al. Paramyxovirus infection in caiman lizards (*Draecena guianensis*). J Vet Diagn Invest 2001;13:143–51.

[114] Foelsch DW, Leloup P. Fatale endemische infection in einem serpentarium. Tierartzl Prax Ausg K Klientiere Heimtiere 1976;4:527–36.

[115] Jacobson ER, Adams HP, Geisbert TW, et al. Pulmonary lesions in experimental ophidian paramyxovirus pneumonia of Aruba Island rattlesnakes, *Crotalus unicolor*. Vet Pathol 1997;34(5):450–9.

[116] Chong HT, Kamarulzaman A, Tan CT, et al. Treatment of acute Nipah encephalitis with ribavirin. Ann Neurol 2001;49:810–3.

[117] Duncan R, Corcoran J, Shou J, et al. Reptilian reovirus: a new fusogenic orthoreovirus species. Virology 2004;319(1):131–40.

[118] Wellehan JFX, Johnson AJ, Roberts JF, et al. Nested PCR amplification and sequencing of a reptile reovirus associated with disease in Mojave rattlesnakes (*Crotalus scutulatus*). Proceedings of the Conference of the American Association of Zoo Veterinarians, San Diego (CA): American Association of Zoo Veterinarians; 2004. p. 644.

[119] Blahak S, Ott I, Vieler E. Comparison of 6 different reoviruses of various reptiles. Vet Res 1995;26:470–6.

[120] Drury SE, Gough RE, Welchman Dde B. Isolation and identification of a reovirus from a lizard, *Uromastyx hardwickii*, in the United Kingdom. Vet Rec 2002;151:637–8.

[121] Jacobson ER. Viruses and viral associated diseases of reptiles. Acta Zool Pathol Antverp 1986;79:73–90.

[122] Ahne W, Thomsen I, Winton J. Isolation of a reovirus from the snake, *Python regius*. Arch Virol 1987;94:135–9.

[123] Blahak S, Gobel T. A case reported of a reovirus infection in an emerald tree boa (*Corallus caninus*). Proc 4th Inter Coll Path Med Rept Amphib, Bad Nauheim: 1991. p. 13–6.

[124] Vieler E, Baumgartner W, Herbst W, et al. Characterization of a reovirus isolate from a rattle snake, *Crotalus viridis*, with neurological dysfunction. Arch Virol 1994;138(3–4): 341–4.

[125] Lamirande EW, Nichols DK, Owens JW, et al. Isolation and experimental transmission of a reovirus pathogenic in ratsnakes (*Elaphe* species). Virus Res 1999;63:135–41.

[126] Marschang RE, Chitty J. Infectious diseases. In: Girling SJ, Raiti P, editors. Manual of reptiles, Gloucester (UK): British Small Animal Veterinary Association; 2004. p. 330–45.

[127] Martin J, Herniou E, Cook J, et al. Human endogenous retrovirus type I-related viruses have an apparently widespread distribution within vertebrates. J Virol 1997;71: 437–43.

[128] Herniou E, Martin J, Miller K, et al. Retroviral diversity and distribution in vertebrates. J Virol 1998;72:5955–66.

[129] Zeigel RF, Clark HF. Electron microscopic observations on a "C"-type virus in cell cultures derived from a tumor-bearing viper. J Natl Cancer Inst 1969;43:1097–102.

[130] Lunger PD, Hardy WD Jr, Clark HF. C-type virus particles in a reptilian tumor. J Natl Cancer Inst 1974;52:1231–5.

[131] Ippen R, Mladenov Z, Konstantinov A. Leukosis with viral presence proven by means of an electron microscope in 2 boa constrictors. Schweiz Arch Tierheilkd 1978;120: 357–68.

[132] Zshiesche W, Konstantinov A, Ippen R, et al. Lymphoid leukemia with presence of C virus particles in a four-lined chicken snake: *Elaphe obsoleta quadrivittata*. [Verhandlungsbericht des 30. Internationalen Symposiums uber die Erkrankungen der Zoo-und Wildtiere. Vehr Int Symp Erkrank Zoot 1988;30:275–8. (in German)

[133] Casey RN, Quackenbush SL, Work TM, et al. Evidence for retrovirus infections in green turtles *Chelonia mydas* from the Hawaiian islands. Dis Aquat Org 1997;31:1–7.

[134] Chandra AMS, Jacobson ER, Munn RJ. Retroviral particles in neoplasms of Burmese pythons (*Python molurus bivittatus*). Vet Pathol 2001;38:561–4.

[135] Carneiro SM, Tanaka H, Kisielius JJ, et al. Occurrence of retrovirus-like particles in various cellular and intercellular compartments of the venom glands from *Bothrops jararacussu*. Res Vet Sci 1992;53:399–401.

[136] Tristem M, Myles T, Hill F. A highly divergent retroviral sequence in the Tuatara (*Sphenodon*). Virology 1995;210:206–11.

[137] Martin J, Kabat P, Herniou E, et al. Characterization and complete nucleotide sequence of an unusual reptilian retrovirus recovered from the Order Crocodylia. J Virol 2002;76: 4651–4.

[138] Schumacher J, Jacobson ER, Homer BL, et al. Inclusion body disease in boid snakes. J Zoo Wildl Med 1994;25:511–24.

[139] Wozniak E, McBride J, DeNardo D, et al. Isolation and characterization of an antigenically distinct 68 kd protein from nonviral intracytoplasmic inclusions in boa constrictors chronically infected with the inclusion body disease virus (IBDV: Retroviridae). Vet Pathol 2000;37:449–59.

[140] Huder JB, Böni J, Hatt JM, et al. Identification and characterization of two closely related unclassifiable endogenous retroviruses in pythons (*Python molurus* and *Python curtus*). J Virol 2002;76:7607–15.

[141] Jacobson ER, Troutman JM, Ginn P, et al. Outbreak of West Nile virus in farmed alligators (*Alligator mississippiensis*). In: Proc Am Assoc Zoo Vet, Minneapolis: 2003. p. 2.

[142] Miller DL, Mauel MJ, Baldwin C, et al. West Nile virus in farmed alligators. Emerg Infect Dis 2003;9:794–9.

[143] Travis D, McNamara T, Glaser A, et al. The national surveillance system for West Nile virus in zoological institutions. In: Proc Am Assoc Zoo Vet, Minneapolis: 2003. p. 265–6.

[144] Steinman A, Banet-Noach C, Tal S, et al. West Nile virus infection in crocodiles. Emerg Infect Dis 2003;9:887–9.

[145] Nir Y, Lasowski Y, Avivi A, et al. Survey for antibodies to arboviruses in the serum of various animals in Israel during 1965–1966. Am J Trop Med Hyg 1969;18:416–22.

[146] Klenk K, Komar N. Poor replication of West Nile virus (New York 1999 strain) in three reptilian and one amphibian species. Am J Trop Med Hyg 2003;69:260–2.

[147] Drury SE, Gough RE, McArthur SD. Detection and isolation of a flavivirus-like agent from a leopard tortoise (*Geochelone pardalis*) in the United Kingdom. Vet Rec 2001;148: 452.

[148] Doi R, Oya A, Shirasaka A, et al. Studies on Japanese encephalitis virus infection of reptiles. II. Role of lizards on hibernation of Japanese encephalitis virus. Jpn J Exp Med 1983;53:125–34.

[149] Shortridge KF, Oya A. Arboviruses. In: Hoff GL, Frye FL, Jacobson ER, editors. Diseases of amphibians and reptiles. New York: Plenum Press; 1984. p. 107–48.

[150] Rehacek J, Nosek J, Gresikova M. Study of the relation of the green lizard (*Lacerta viridis* Laur.) to natural foci of tick-borne encephalitis. J Hyg Epidemiol Microbiol Immunol 1961; 5:366–72.

[151] Sekeyova M, Gresikova M, Lesko J. Formation of antibody to tick-borne encephalitis virus in *Lacerta viridis* and *L. agilis* lizards. Acta Virol 1970;14:87.

[152] Hayes RO, Daniels JB, Maxfield HK, et al. Field and laboratory studies on eastern equine encephalitis in warm- and cold-blooded vertebrates. Am J Trop Med Hyg 1964;13:595–606.

[153] Bowen GS. Prolonged western equine encephalitis viremia in the Texas tortoise (*Gopherus berlandieri*). Am J Trop Med Hyg 1977;26:171–5.

[154] Thomas LA, Patzer ER, Cory JC, et al. Antibody development in garter snakes (*Thamnophis* spp.) experimentally infected with western equine encephalitis virus. Am J Trop Med Hyg 1980;29:112–7.

[155] Hoff GL, Trainer DO. Arboviruses in reptiles: isolation of a bunyamwera group virus from a naturally infected turtle. J Herpetol 1973;7:55–62.

[156] Smith AW, Anderson MP, Skilling DE, et al. First isolation of calicivirus from reptiles and amphibians. Am J Vet Res 1968;47:1718–21.

[157] Barlough JE, Matson DO, Skilling DE, et al. Isolation of reptilian calicivirus crotalus type 1 from feral pinnipeds. J Wildl Dis 1998;34:451–6.

[158] Matson DO, Berke T, Dinulos MB, et al. Partial characterization of the genome of nine animal caliciviruses. Arch Virol 1996;141:2443–56.

[159] Cropp CB. Reptilian rhabdoviruses. In: Hoff GL, Frye FL, Jacobson ER, editors. Diseases of amphibians and reptiles. New York: Plenum Press; 1984. p. 149–57.

[160] Cook RS, Trainer DO, Glazener WC, et al. A serological study of infectious diseases of wild populations in south Texas. Trans N Am Wildl Nat Res Conf 1965;30:142–55.

ELSEVIER
SAUNDERS

VETERINARY
CLINICS
Exotic Animal Practice

Vet Clin Exot Anim 8 (2005) 53–65

Amphibian virology

April J. Johnson, DVM*,
James F.X. Wellehan, DVM, MS

College of Veterinary Medicine, University of Florida, Gainesville, FL 32610, USA

Amphibians are a diverse group of species; much work remains to be done to elucidate the viruses of amphibians. In addition to frank viral diseases, viral infections may play a role in the establishment of bacterial, fungal, and parasitic diseases and are an underlying cause of neoplasia. It is important for the amphibian clinician to recognize disease syndromes and pathology that are consistent with viral etiology, to understand viral diagnostic testing, and to have good communication with an experienced amphibian pathologist and a good viral diagnostic lab.

Diagnostic tests

Diagnostic tests for the viruses of amphibians may be divided into two categories—testing for immune response to virus and testing for the presence of virus. It is important for the clinician to be aware of the information that is given by the test and to interpret results appropriately.

Tests that look for the presence of immune response to virus include serum neutralization testing, ELISAs, hemagglutination inhibition, agarose gel immunodiffusion testing, and T-cell proliferation assays. It is important to consider what aspect of immune response is being tested. T-cell proliferation assays test for the presence of a cellular immune response; the others test for the presence of a humoral immune response. A humoral immune response primarily produces antibodies against pathogens, whereas a cellular immune response primarily uses cytotoxic T cells to trigger self-destruction of host cells that are infected with pathogens. Typically, a cellular immune response is most protective for viruses. Often, but not always, cellular and humoral immune

---

* Corresponding author.
E-mail address: johnsona@mail.vetmed.ufl.edu (A.J. Johnson).

doi:10.1016/j.cvex.2004.09.001

responses are found together. An animal may have a cellular immune response without a significant humoral immune response, and vice versa.

It also is important to consider whether the patient has had time to develop an immune response to a pathogen. In acute infections, a specific immune response may not be present. There are large seasonal and temperature effects on the immune response of amphibians, as well as species differences [1–3]. Other factors, including metamorphosis, hibernation, and exogenous corticosteroid administration, also may affect development of a measurable immune response [4–6].

Virus isolation or polymerase chain reaction (PCR) will give information regarding the presence of virus. These tests require virus in the sample that is submitted. The selection of sample for these tests is highly important. For example, in a patient that has a viral infection that is not viremic at the time of sample collection, test results on blood will be negative. This does not mean that there is not virus elsewhere in the patient. For virus isolation, the virus needs to be viable, so the sample needs to be collected and stored in a manner so that the virus is not inactivated. PCR testing requires intact nucleic acid, but does not require live virus.

PCR testing is highly sensitive; this makes it susceptible to false positives that result from contamination. A PCR involves use of primers that are specific for a nucleic acid (DNA or RNA) sequence and repeated use of an enzyme to synthesize copies of the nucleic acid sequence between them. Primers may be designed to be highly specific for an organism or to amplify nucleic acid from any organism in a given group. A PCR that uses primers that are designed to amplify nucleic acid from multiple organisms is called a consensus PCR. After nucleic acid has been amplified by PCR, the identity of the product needs to be validated. The least specific method of validating the PCR product is gel electrophoresis, which shows that the product is the correct length. To increase the specificity, the product may be digested by restriction enzymes, which recognize short sequences and cut them. This shows that the PCR product contains restriction enzyme sequences that are found within the expected product. A more specific method of validating the identity of the PCR product is to hybridize the product with a labeled nucleic acid probe from the expected product. The most specific method is to sequence the PCR product. A good diagnostic laboratory occasionally will sequence PCR products to ensure quality control. This should be available—for an additional cost—as a validation to any clinician who submits a sample for PCR. Any consensus PCR product should be identified by DNA sequencing.

If the presence of virus is identified by virus isolation, electron microscopy, or nonspecies-specific DNA in situ hybridization, it is recommended strongly that the virus be identified further. Identification of viral types and species provide useful diagnostic, prognostic, and epidemiologic information for the clinician. Significant clinical differences may exist between viruses in the same family from the same host species and between different host species with the same virus. In primates, this is illustrated by the clinical difference between infection with human herpesvirus 1 (herpes simplex), which causes cold sores in humans, and the closely related cercopithecine herpesvirus 1 (herpes B), which causes cold sores in macaques and a fatal encephalitis in humans. In gibbons or marmosets, human herpesvirus 1 causes a fatal encephalitis.

Regardless of the type of diagnostic testing for virus, this information needs to be correlated with clinical and histologic findings in the animal. Neither the presence of an immune response nor the presence of virus necessarily indicate the presence of viral disease. Virus infection may occur without clinical disease and the presence of an immune response does not mean that there is current infection.

Control

When managing viral diseases in the individual animal, it is important for the clinician to look at the whole picture. Typically, suboptimal husbandry is a large contributing factor to the onset of disease; correction of husbandry deficits is one of the most important aspects of treatment. Diet, lack of an even and optimal thermal gradient so that the animal can thermoregulate properly, water quality, overcrowding, lack of appropriate hiding/perching areas, overhandling, inappropriate humidity, poor sanitary conditions, and other environmental stressors play important roles in the onset of disease. Supportive care for the sick animal should be provided, including fluid therapy, treatment of bacterial or fungal secondary infections, and oxygen therapy if there is respiratory compromise. Amphibians have been shown to select higher temperatures in response to bacterial infection [7]; an ambulatory patient is better able to select an appropriate temperature than the clinician. Moribund animals may be unable to move to thermoregulate appropriately; therefore, the clinician must be even more attentive to thermal issues in these animals.

Amphibians produce and secrete antimicrobial peptides from their skin, which have been shown to have antibacterial and antifungal activity [8]. Recently, several antimicrobial peptide isolates have demonstrated antiviral activity against a herpesvirus and a *Ranavirus* experimentally [9,10]. The natural occurrence of virus protection as a result of antimicrobial peptides has not been determined. Specific antiviral drugs (eg, acyclovir for herpesviruses and iridoviruses or cidofovir for adenoviruses) are discussed later; however, it is important to recognize that no pharmacokinetic, efficacy, or safety studies have been done for any of these drugs in any amphibian species.

When dealing with an amphibian collection, it is important to know the infection status of the collection for the pathogen of concern. When introducing new animals, they should be tested for diseases of concern, quarantined for a minimum of 90 days, and tested again before leaving quarantine. Many viruses are host-adapted and cause mild to subclinical disease in their natural host species under optimal husbandry conditions.

When an aberrant host species is infected, however, serious clinical disease may result. Therefore, from an infectious disease standpoint, mixed species enclosures should be avoided. Selection of sources, if acquiring new animals, must be considered. When an animal is caught from the wild, a large amount of stress is involved. Typically, these animals are held in overcrowded, inappropriate conditions for what may be a long period of time until enough are collected to be shipped to an amphibian importer. They are usually brought in without quarantine and a large amount of cross-contamination occurs with diverse amphibian species from all over the world. An amphibian importer is an excellent source of fresh material for the virologist, but less than an ideal source for acquisition of animals for a collection. Captive bred animals, preferably from a breeder who works with a single species, are preferred. Another factor that plays a role in the development of disease is genetic diversity. Outbred, wild-type animals have a more robust immune system [11]. Typically, inbred color mutants are more susceptible to disease which makes them less desirable in a collection. Conversely, maintenance of a few color mutant animals—as disease sentinels—in a collection may be useful. In general, a closed collection with careful addition of new animals only for maintenance of genetic diversity is optimal.

If an animal is diagnosed with viral disease, the value of the collection and of the animal, as well as the transmissibility of the disease, must be weighed. Enveloped viruses (eg, herpesviruses) tend to be less stable in the environment; prevention of transmission from one enclosure to another by human vectors is feasible with good hygiene practices, including not transferring food dishes or other cage furniture between enclosures, washing hands between enclosures, maintaining clean enclosures, using foot baths, and dealing with sick animals last. Conversely, nonenveloped viruses (eg, adenoviruses, iridoviruses) tend to be stable in the environment; preventing spread presents a greater challenge. Moving affected and exposed animals to a separate facility or culling should be considered.

Use of vaccines for prevention of viral infection in amphibians is an area that remains to be explored. In fish, an iridovirus vaccine was shown to be effective [12]. When considering vaccine design for an intracellular pathogen (eg, virus), it is important to elicit a cellular immune response. To activate $CD8^+$ T cells, the viral antigen needs to be presented by major histocompatibility complex (MHC) class I molecules. These molecules are expressed on the surface of most cells and present peptides that are found within the cell. Therefore, to elicit an optimal cellular immune response, a viral vaccine should be able to enter cells. Killed vaccines do not do this. Another option is the use of a modified live vaccine that has been passed through cell culture until it loses virulence. The main concern here is the possibility of return to virulence; this has occurred with modified live vaccines [13]. Recombinant vaccines that use an unrelated avirulent virus which expresses an immunogenic peptide from the pathogenic virus of concern, may hold the most promise for future vaccine development.

DNA viruses

The DNA viruses tend to have larger genomes. DNA viruses mutate more slowly than RNA viruses. Many DNA viruses have evolutionary histories that parallel those of their hosts [14,15]. This can provide interesting information about host as well as viral evolution, and makes consensus PCR primer design easier. DNA viruses that have been identified in amphibians include herpesviruses, iridoviruses, adenoviruses, and parvoviruses.

Herpesviruses

Herpesviruses are large, enveloped, double-stranded, linear DNA viruses that persistently infect their host. Amphibian herpesviruses are phylogenetically different from mammalian, avian, and reptilian herpesviruses. Lucké tumor herpesvirus or ranid herpesvirus (RaHV)-1 is related most closely to ictalurid herpesvirus 1, a channel catfish virus that essentially is unrelated to mammalian herpesviruses [16] and is not categorized as an α-, β-, or γ-herpesvirus by the International Committee on the Taxonomy of Viruses. RaHV-1 is a causative agent of renal adenocarcinomas in the leopard frog (*Rana pipiens*) [17]. This disease was recognized first in 1934 [18] and is characterized by the presence of intranuclear inclusion bodies in tumor cells (Figs. 1 and 2). Virus replication is temperature dependent and occurs at

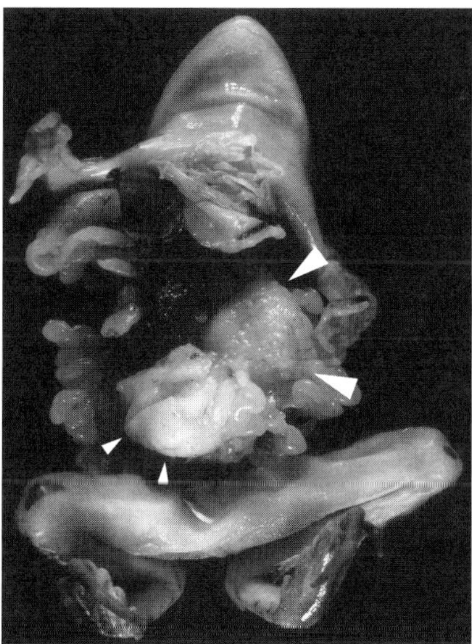

Fig. 1. Photograph of renal adenocarcinoma in a leopard frog that commonly is associated with a herpesvirus. Large arrows depict carcinoma originating from the cranial pole of the kidney. Small arrows depict tumor invasion of adjacent oviduct. (Courtesy of Brian Stacy, DVM, Dipl ACVP, Gainesville, Florida.)

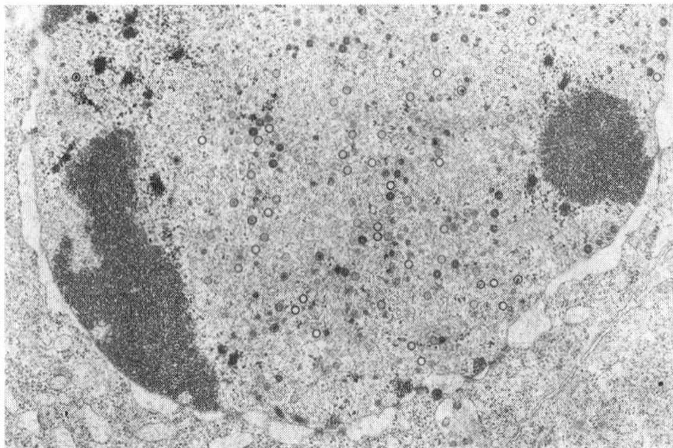

Fig. 2. Electron micrograph of ranid herpesvirus particles in a nucleus from a leopard frog with renal adenocarcinoma. (Courtesy of Robert McKinnell, PhD, St. Paul, Minnesota.)

a higher rate at lower temperatures [19]. Tumors often develop in infected tadpoles at metamorphosis [20]. A second herpesvirus, frog virus 4 (also called RaHV-2), was isolated from urine of leopard frogs who had adenocarcinomas. Transmission to other frogs failed to result in tumor development which suggested that it was a different virus [21]. A third anuran herpesvirus was detected by electron microscopy in skin lesions of spring frogs, *R dalmatina*, but it was not isolated or further characterized [22]. A herpesvirus like agent also was seen on electron microscopy in skin papillomas of the Japanese newt (*Cynops pyrrhogaster*) [23,24]. Consensus primers that are used commonly to detect a portion of the DNA-dependent DNA polymerase in mammalian, avian, and reptilian herpesviruses [25] cannot be used to detect RaHV-1 or -2. No serologic or PCR diagnostic tests are available for amphibian herpesvirus at this time.

Iridoviruses

Iridoviruses are large, nonenveloped, double-stranded, cytoplasmic DNA viruses [26] that infect invertebrates and poikilothermic vertebrates. The family *Iridoviridae* consists of four genera. *Chloriridovirus* and *Iridovirus* infect insects, *Lymphocystivirus* infects fish, and *Ranavirus* is capable of infecting fish, amphibians, and reptiles [27]. Frog virus 3, the type species for the genus *Ranavirus*, was isolated first from a renal carcinoma in a leopard frog [28], although it was later determined that there was no association between the virus and the tumor [29]. Shortly thereafter, a *Ranavirus* was recovered from edematous bullfrog tadpoles and was named tadpole edema virus [30]. Significant research with amphibian iridoviruses did not make much progress until the early 1990s, when worldwide declines in amphibians spurred new interest in the role of these viruses in mortality events. Many viruses from the genus *Ranavirus* have been identified from die-offs virtually worldwide [30–36]. In a study of 64 amphibian mortality and morbidity events in the United States, iridovirus was the most common cause of mortality [32]. Late larval forms were most susceptible and epizootics were associated with increased population densities. Adult amphibians and larval stages can be asymptomatic carriers [30,37] and certain iridoviruses may be nonhost-specific as demonstrated naturally [38] and experimentally [30,39,40]; sympatric species of fish and amphibians were susceptible to the same virus. Transmission can occur by ingestion of infected tissue, infected water, and by direct contact [31].

Clinical signs range from none to severe. Death was observed in experimental infections without any external signs of disease [34,37,41]. Clinical signs that have been associated with natural or experimental infection with iridovirus include lethargy, anorexia, buoyancy problems, swimming in circles, loss of righting ability, reddened or swollen areas near the gills and hind limbs, pale raised foci in the skin, and cutaneous erosions or ulcers [30,31,33,34,36–38,41]. Infected anurans can present with "red-leg" and it was suggested that iridovirus may be the primary causative agent, whereas bacterial infections, typically associated with "red-leg", are secondary [36]. Infected salamanders can produce thick mucus that adheres to the skin and have loose feces, bloody stools, and vomiting [31,34]. Jankovich et al [31] experimentally demonstrated a progression of clinical signs in salamanders that was temperature dependent. Signs started with small white polyp formations that spread to cover almost the entire body and developed into hemorrhaging of the epidermis.

Subcutaneous edema, ascites, and petechiation of serosal surfaces of the gastrointestinal tract often can be seen at necropsy [30,33,34]. Iridovirus can cause necrosis of hematopoietic tissue in the kidney, spleen, and liver [37]. Focal areas of necrosis have been seen in the liver, kidney, spleen, epidermis, stomach, intestines, oral cavity, lung, pancreas, fat bodies, bladder, and thymus [30,33,34,36,37,41]. Iridovirus forms basophilic intracytoplasmic inclusion bodies that are found in most of the same tissues [34].

Frog erythrocytic virus (FEV) is an iridovirus-like organism that has been documented in erythrocytes of several species of frogs and toads [42–46]. Initially, it was believed to be a protozoan (*Toddia* or *Pirhemocyton* spp); however, ultrastructural and biochemical characterization demonstrated that the virus like particles were most consistent with iridoviruses [47]. FEV has not been isolated in cell culture and no nucleotide sequence exists to characterize it further. The clinical significance of this virus has not been determined.

Diagnosis of *Ranavirus* is made with a combination of clinical signs, characteristic histologic lesions, virus isolation, PCR, and transmission electron microscopy. Sequencing of the major capsid protein gene often is used to confirm iridovirus. PCR and sequencing is performed in the laboratory of Dr. Elliott Jacobson at the University of Florida. Diagnosis of FEV is made by detecting inclusion bodies in erythrocytes under light microscopy followed by electron microscopy to detect virus particles. Because iridoviruses have a viral thymidine kinase—the enzyme that converts acyclovir and related drugs to their active form—it is possible that acyclovir may be effective against iridoviruses. No studies have been done to evaluate this efficacy. In fish, an iridovirus vaccine was shown to be effective [12], but no vaccination studies have been done in amphibians.

Adenoviruses

Adenoviruses are double-stranded, nonenveloped DNA viruses. All characterized amphibian adenoviruses, as well as with some bird adenoviruses [49], are in the genus *Siadenovirus* [48]. Frog adenovirus (FrAdV)-1 was isolated first from a renal tumor in a leopard frog (*R pipiens*) [50] and later from a second leopard frog tumor [51]. Koch's postulates were not fulfilled when transmission studies of tadpoles and chick embryos failed to result in pathology; however, adenovirus was isolated consistently from the renal tumor and was not isolated from other nontumor tissue. The complete genome of FrAdV-1 was sequenced and is related most closely to turkey adenovirus (also *Siadenovirus*) [52]. The genome is the smallest of all adenovirus genomes at 26,163 base pairs. Another report of adenovirus-like particles that were cultured from a common frog (*R temporaria*) [36] was not characterized further and the virus was believed to be an incidental finding.

Histologically, adenoviruses often cause intranuclear inclusion bodies that distend the nucleus. Few attempts have been made at antemortem diagnosis of adenoviral infection in amphibians; it is likely that increased surveillance would result in the recognition of additional disease syndromes and associated adenoviral types. Adenoviral consensus PCR and sequencing is available from the laboratory of Dr. Elliott Jacobson at the University of Florida. The use of certain antiviral drugs (eg, cidofovir) has been shown to be effective for therapy of some adenoviral diseases [53]; this also may prove to be the case with some amphibian adenoviruses.

Parvoviruses

Parvoviruses are single-stranded, nonenveloped DNA viruses. An inclusion body myositis was observed in the tongue of two spring peepers (*Pseudacris crucifer*) in which virus-like particles, which were consistent in size, shape, and location with parvoviruses, were observed within inclusions [54]. Virus isolation was not attempted and no molecular diagnostics were performed to confirm that the particles were parvovirus particles. Transmission studies were not performed to confirm that these were the causative agents of the observed clinical signs. Amphibian parvoviruses have not been isolated successfully; no serologic or PCR diagnostics are available to the clinician for amphibian parvoviruses.

Poxviruses

Initial reports of poxvirus like particles in the skin of amphibians were later determined to be melanosomes [36]. No poxviruses have been found in amphibians.

RNA viruses

The RNA viruses tend to have smaller genomes. RNA viruses mutate more rapidly than DNA viruses. Many RNA viruses mutate so rapidly that evolutionary change of the virus can be followed within a single host [55]. This makes consensus PCR primer design more difficult. RNA viruses that have been identified in amphibians include retroviruses, flaviviruses, togaviruses, and caliciviruses.

Retroviruses

Retroviruses are single-stranded, positive-sense, enveloped RNA viruses that contain a DNA step in their replication. They are divided into seven genera. Sequences that were obtained from the amphibian retroviruses cluster with the *Spumavirus*, *Epsilonretrovirus*, and *Gammaretrovirus* genera [56–58]. Endogenous retroviruses have been sequenced from each of the three orders of amphibians and include caecilians, salamanders, newts, frogs, and toads [57]. No disease syndrome was associated with any of these retroviruses. C-type retroviral particles were observed in tumor cells of the black-spotted frog, *R nigromaculata*, at a laboratory in Japan. A high prevalence of pancreatic carcinomas developed in the population over 5 years; however, transmission studies were not performed to confirm that the retrovirus was the causative agent and no sequence was obtained for the retrovirus [59]. No serologic or PCR diagnostics are available to the clinician for amphibian retroviruses.

Flaviviruses

Flaviviruses are small, enveloped, unsegmented, positive-sense, single-stranded RNA viruses that typically are borne by arthropods. As a group, flaviviruses include several serious diseases in humans, including West Nile virus, Japanese encephalitis, hepatitis C, yellow fever, and dengue fever. All members of the family *Flaviviridae*, for which there is evidence of infection in amphibians, are in the genus *Flavivirus*.

Experimental infection of 24 American bullfrogs (*R catesbeiana*) with West Nile virus resulted in viremia in 2 frogs [60]. The marsh frog (*R ridibunda*) was shown to be a potential reservoir for West Nile virus in Russia [61]. Neutralizing antibodies to St. Louis encephalitis virus have been found in a leopard frog (*R pipiens*) [62]. Black-spotted frogs (*R nigromaculata*) have been infected experimentally with Japanese encephalitis virus [63] and there is serologic evidence of Japanese encephalitis infection in tiger frogs (*R tigrina*)

[64]. Overall, zoonotic flaviviruses have not been found to be a serious cause of disease in amphibians. Although it does not appear to be of clinical concern to the amphibian patient, PCR for West Nile virus is available in several laboratories.

Togaviruses

Togaviruses are small, enveloped, unsegmented, positive-sense, single-stranded RNA viruses that typically are borne by arthropods. As a group, togaviruses include several serious diseases in humans, including eastern equine encephalitis, western equine encephalitis, Venezuelan equine encephalitis, and rubella. All members of the family *Togaviridae*, for which there is evidence of infection in amphibians, are in the genus *Alphavirus*.

Sindbis virus has been isolated from the marsh frog (*R ridibunda*) in Slovakia [65], although lesions or clinical signs were not described. There is serologic evidence for western equine encephalitis infection of leopard frogs (*R pipiens*) in Canada [66]. One of five leopard frogs that were infected experimentally with eastern equine encephalitis virus died 24 hours post-infection and was found to be viremic. The other four leopard frogs, as well as five green frogs (*R clamitans*) and an American bullfrog (*R catesbeiana*) that were infected experimentally were not viremic 60 days postinfection [67]. Overall, zoonotic togaviruses have not been found to be a major cause of disease in amphibians.

Caliciviruses

Caliciviruses are small, nonenveloped, positive-sense, single-stranded linear RNA viruses. Caliciviruses have been isolated from two Bell's horned frogs (*Ceratophrys ornata*) [68]. The calicivirus that was identified from amphibians is in the genus *Vesivirus*. Although both frogs had interstitial pneumonia, no specific disease syndrome was associated with the virus and no disease was seen in snakes or pigs that were infected experimentally. No serologic or PCR diagnostics are available to the clinician for amphibian caliciviruses.

References

[1] Zapata AG, Varas A, Torroba M. Seasonal variations in the immune system of lower vertebrates. Immunol Today 1992;13:142–7.

[2] Cone RE, Marchalonis JJ. Cellular and humoral aspects of the influence of environmental temperature on the immune response of poikilothermic vertebrates. J Immunol 1972;108:952–7.

[3] Lin HH, Rowlands DT Jr. Thermal regulation of the immune response in South American toads (*Bufo marinus*). Immunol 1973;24:129–33.

[4] Cooper EL, Wright RK, Klempau AE, et al. Hibernation alters the frog's immune system. Cryobiology 1992;29:616–31.

[5] Ruben LN, Clothier RH, Murphy GL, et al. Thyroid function and immune reactivity during metamorphosis in *Xenopus laevis*, the South African clawed toad. Gen Comp Endocrinol 1989;76:128–38.

[6] Tournefier A. Corticosteroid action on lymphocyte subpopulations and humoral immune response of axolotl (urodele amphibian). Immunol 1982;46:155–62.

[7] Muchlinski AE. The energetic cost of the fever response in three species of ectothermic vertebrates. Comp Biochem Physiol A 1985;81:577–9.

[8] Rollins-Smith LA, Doersam JK, Longcore JE, et al. Antimicrobial peptide defenses against pathogens associated with global amphibian declines. Dev Comp Immunol 2002;26:63–72.

[9] Chinchar VG, Wang J, Murti G, et al. Inactivation of frog virus 3 and channel catfish virus by esculentin-2P and ranateurin-2P, two antimicrobial peptides isolated from frog skin. Virology 2001;288:351–7.

[10] Chinchar VG, Bryan L, Silphadaung U, et al. Inactivation of viruses infecting ectothermic animals by amphibian and piscine antimicrobial peptides. Virology 2004;323:268–75.

[11] McClelland EE, Penn DJ, Potts WK. Major histocompatibility complex heterozygote superiority during coinfection. Infect Immun 2003;71:2079–86.

[12] Opriessnig T, Halbur PG, Yoon KJ, et al. Comparison of molecular and biological characteristics of a modified live porcine reproductive and respiratory syndrome virus (PRRSV) vaccine (ingelvac PRRS MLV), the parent strain of the vaccine (ATCC VR2332), ATCC VR2385, and two recent field isolates of PRRSV. J Virol 2002;76:11837–44.

[13] Nakajima K, Maeno Y, Honda A, et al. Effectiveness of a vaccine against red sea bream iridoviral disease in a field trial test. Dis Aquat Org 1999;36:73–5.

[14] McGeoch DJ, Davison AJ. The molecular evolutionary history of the herpesviruses. In: Domingo E, Webster R, Holland J, editors. Origin and evolution of viruses. San Diego (CA): Academic Press; 1999. p. 441–65.

[15] Wellehan JFX, Jarchow JL, Reggiardo C, et al. A novel herpesvirus associated with hepatic necrosis in a San Esteban chuckwalla, *Sauromalus varius*. J Herpetol Med Surg 2003;13(3): 15–9.

[16] Davison AJ. Channel catfish virus: a new type of herpesvirus. Virology 1992;186:9–14.

[17] Naegele RF, Granoff A, Darlington RW. The presence of the Lucké herpesvirus genome in induced tadpole tumors and its oncogenicity: Koch-Henle postulates fulfilled. Proc Natl Acad Sci USA 1974;71:830–4.

[18] Lucké B. A neoplastic disease of the kidney of the frog, *Rana pipiens*. Am J Cancer 1934;20: 352–79.

[19] Carlson DL, Rollins-Smith LA, McKinnell RG. The Lucké herpesvirus genome: its presence in neoplastic and normal kidney tissue. J Comp Pathol 1994;110:349–55.

[20] Tweedell KS. Induced oncogenesis in developing frog kidney cells. Cancer Res 1967;27: 2042–52.

[21] Tweedell KS. Herpesviruses: interaction with frog renal cells. In: Ahne W, Kurstak E, editors. Viruses of lower vertebrates. Heidelberg (Germany): Springer Verlag; 1989. p. 13–29.

[22] Bennati R, Bonetti M, Lavazza A, et al. Skin lesions associated with herpesvirus-like particles in frogs (*Rana dalmatina*). Vet Rec 1994;135:625–6.

[23] Asashima M, Komazaki S, Satou C, et al. Seasonal and geographical changes of spontaneous skin papillomas in the Japanese newt *Cynops pyrrhogaster*. Cancer Res 1982; 42(9):3741–6.

[24] Pfeiffer CJ, Asashima M, Hirayasu K. Ultrastructural characterization of the spontaneous papilloma of Japanese newts. J Submicrosc Cytol Pathol 1989;21(4):659–68.

[25] VanDevanter DR, Warrener P, Bennett L, et al. Detection and analysis of diverse herpesviral species by consensus primer PCR. J Clin Microbiol 1996;34:1666–71.

[26] Williams T. The iridoviruses. Adv Virus Res 1996;46:345–412.

[27] Mao J, Hedrick RP, Chinchar VG. Molecular characterization, sequence analysis and taxonomic position of newly isolated fish iridoviruses. Virology 1997;229:212–20.

[28] Granoff A, Came PE, Rafferty KA Jr. The isolation and properties of viruses from *Rana pipiens*: their possible relationship to the renal adenocarcinoma of the leopard frog. Ann N Y Acad Sci 1965;126:237–55.

[29] Granoff A, Came PE, Breeze DC. Viruses and renal adenocarcinoma of *Rana pipiens*. I. The isolation and properties of virus from normal and tumor tissue. Virology 1966;29:133–48.

[30] Wolf K, Bullock GL, Dunbar CE, et al. Tadpole edema virus: a viscerotrophic pathogen for anuran amphibians. J Infect Dis 1968;118:253–62.

[31] Jankovich JK, Davidson EW, Morado JF, et al. Isolation of a lethal virus from the endangered tiger salamander *Ambystoma tigrinum stebbinsi*. Dis Aquat Org 1997;31:161–7.

[32] Green DE, Converse KA, Schrader AK. Epizootiology of sixty-four amphibian morbidity and mortality events in the USA, 1996–2001. Ann N Y Acad Sci 2002;969:323–39.

[33] Docherty D, Meteyer CU, Wang J, et al. Diagnostic and molecular evaluation of three iridovirus-associated salamander mortality events. J Wildl Dis 2003;39:556–66.

[34] Bollinger TK, Mao J, Schock D, et al. Pathology, isolation and preliminary molecular characterization of a novel iridovirus from tiger salamanders in Saskatchewan. J Wildl Dis 1999;35:413–29.

[35] Speare R, Smith JR. An iridovirus-like agent isolated from the ornate burrowing frog *Limnodynastes ornatus* in northern Australia. Dis Aquat Org 1992;14:51–7.

[36] Cunningham AA, Langton TES, Bennett PM, et al. Pathological and microbiological findings from incidents of unusual mortality of the common frog (*Rana temporaria*). Philos Trans R Lond B Biol Sci 1996;351:1529–57.

[37] Cullen BR, Owens L. Experimental challenge and clinical cases of Bohle iridovirus (BIV) in native Australian anurans. Dis Aquat Org 2002;49:83–92.

[38] Mao J, Green DE, Fellers G, et al. Molecular characterization of iridoviruses isolated from sympatric amphibians and fish. Virus Res 1999;63:45–52.

[39] Clark JF, Gray G, Fabian F, et al. Comparative studies of amphibian cytoplasmic virus strains isolated from the leopard frog, bullfrog, and newt. In: Mizell A, editor. Biology of amphibian tumors. Recent results in cancer research. New York: Springer-Verlag; 1969. p. 310–26.

[40] Moody NJG, Owens L. Experimental demonstration of the pathogenicity of a frog virus, Bohle iridovirus, for a fish species barramundi *Lates calcarifer*. Dis Aquat Org 1994;18: 95–102.

[41] Cullen BR, Owens L, Whittington RJ. Experimental infection of Australian anurans (*Limnodynastes terraereginae* and *Litoria latopalmata*) with Bohle iridovirus. Dis Aquat Org 1995;23:83–92.

[42] Bernard GW, Cooper EL, Mandell ML. Lamellar membrane encircled viruses in the erythrocytes of *Rana pipiens*. J Ultrastruct Res 1968;26:8–16.

[43] Desser SS, Barta JR. An intraerythrocytic virus and rickettsia of frogs from Algonquin Park, Ontario. Can J Zool 1984;62:1521–4.

[44] Gruia-Gray J, Ringuette M, Desser SS. Cytoplasmic localization of the DNA virus frog erythrocytic virus. Intervirology 1992;33:159–64.

[45] Alves de Matos AP, Paperna I, Lainson R. An erythrocytic virus of the Brazilian tree-frog, *Phrynohyas venulosa*. Mem Inst Oswaldo Cruz 1995;90:653–5.

[46] Werner JK. Blood parasites of amphibians from Sichuan Province, People's Republic of China. J Parasitol 1993;79:356–63.

[47] Gruia-Gray J, Petric M, Desser S. Ultrastructural, biochemical and biophysical properties of an erythrocytic virus of frogs from Ontario, Canada. J Wildl Dis 1989;25:497–506.

[48] Davison AJ, Harrach B. Siadenovirus. In: Tidona CA, Darai G, editors. The Springer index of viruses. New York: Springer-Verlag; 2002. p. 29–33.

[49] Pitcovski J, Mualem M, Rei-Koren Z, et al. The complete DNA sequence and genome organization of the avian adenovirus, hemorrhagic enteritis virus. Virology 1998;249:307–15.

[50] Clark HF, Michalski F, Tweedell KS, et al. An adenovirus, FAV-1, isolated from the kidney of a frog (*Rana pipiens*). Virology 1973;51:392–400.

[51] Wong WY, Tweedell KS. Two viruses from the Lucke tumor isolated in a frog pronephric cell line. Proc Soc Exp Biol Med 1974;145(4):1202–6.

[52] Davison AJ, Wright KM, Harrach B. DNA sequence of frog adenovirus. J Gen Virol 2000; 81:2431–9.

[53] Romanowski EG, Yates KA, Gordon YJ. Antiviral prophylaxis with twice daily topical cidofovir protects against challenge in the adenovirus type 5/New Zealand rabbit ocular model. Antiviral Res 2001;52:275–80.

[54] Raymond JT, Reichard T, Shellabarger W, et al. Inclusion body myositis in spring peepers (*Pseudacris crucifer*). J Vet Diagn Invest 2002;14:501–3.

[55] Moya A, Holmes EC, Gonzalez-Candelas F. The population genetics and evolutionary epidemiology of RNA viruses. Nat Rev Microbiol 2004;2:279–88.

[56] Tristem M, Herniou E, Summers K, et al. Three retroviral sequences in amphibians are distinct from those in mammals and birds. J Virol 1996;70:4864–70.

[57] Herniou E, Martin J, Miller K, et al. Retroviral diversity and distribution in vertebrates. J Virol 1998;72:5955–66.

[58] Kambol R, Kabat P, Tristem M. Complete nucleotide sequence of an endogenous retrovirus from the amphibian, *Xenopus laevis*. Virology 2002;311:1–6.

[59] Masahito P, Nishioka M, Ueda H, et al. Frequent development of pancreatic carcinomas in the *Rana nigromaculata* group. Cancer Res 1995;55:3781–4.

[60] Klenk K, Komar N. Poor replication of West Nile virus (New York 1999 strain) in three reptilian and one amphibian species. Am J Trop Med Hyg 2003;69:260–2.

[61] Kostiukov MA, Alekseev AN, Bulychev VP, et al. Experimental evidence for infection of Culex pipiens L. mosquitoes by West Nile fever virus from *Rana ridibunda* pallas and its transmission by bites. Med Parazitol (Mosk) 1986;6:76–8.

[62] Whitney E, Jamnback H, Means RG, et al. Arthropod-borne-virus survey in St. Lawrence County, New York. Arbovirus reactivity in serum from amphibians, reptiles, birds, and mammals. Am J Trop Med Hyg 1968;17:645–50.

[63] Kawasaki M. Hemagglutination inhibiting substances in Japanese encephalitis infected frogs during and after hibernation. Igaku To Seibutsugaku 1972;84:113–6.

[64] Shortridge KF, Oya A, Kobayashi M, et al. Japanese encephalitis virus antibody in cold-blooded animals. Trans R Soc Trop Med Hyg 1977;71:261–2.

[65] Kozuch O, Labuda M, Nosek J. Isolation of sindbis virus from the frog *Rana ridibunda*. Acta Virol 1978;22:78.

[66] Burton AN, McLintock J, Rempel JG. Western equine encephalitis virus in Saskatchewan garter snakes and leopard frogs. Science 1966;154:1029–31.

[67] Hayes RO, Daniels JB, Maxfield HK, et al. Field and laboratory studies on eastern equine encephalitis in warm- and cold-blooded vertebrates. Am J Trop Med Hyg 1964;13:595–606.

[68] Smith AW, Anderson MP, Skilling DE, et al. First isolation of calicivirus from reptiles and amphibians. Am J Vet Res 1986;47:1718–21.

ELSEVIER
SAUNDERS

Vet Clin Exot Anim 8 (2005) 67–84

VETERINARY
CLINICS
Exotic Animal Practice

Viruses of pet fish

Barbara D. Petty, DVM[a],*, William A. Fraser, BS[b]

[a]*Department of Large Animal Clinical Sciences, University of Florida,
7922 NW 71 Street, Gainesville, FL 32653, USA*
[b]*Animal Disease Diagnostic Laboratory, Division of Animal Industry, Department of
Agriculture and Consumer Services, P.O. Box 458006, Kissimmee, FL 34745, USA*

Many veterinarians who work with pet fish may see viruses infrequently because adverse environmental conditions are responsible for most fish health problems. It is important, however, for veterinarians to be cognizant of the few viruses of ornamental fish that have serious implications. This article is a guide to the most common viruses that affect fish that are kept as pets or display animals. Some of the viruses that are included here are presented simply as case reports, given that the incidence and confirmation of viral disease in pet fishes occur sporadically. Where more information is available, clinical findings are discussed in more detail.

Although several fish cell lines are available, many viruses of pet fishes will not grow in them. Diagnosis of such viruses is limited to electron microscopic examination of tissues. Other diagnostic methods, such as immunofluorescence, ELISA, and virus neutralization, are not available for many of the viral diseases that are mentioned here, but some laboratories may be able to provide such tests. Polymerase chain reaction (PCR) methods for several fish viruses are available; for some viruses (eg, koi herpesvirus), PCR may be a more reliable detection method than virus isolation or electron microscopy.

Viruses that affect many species

Iridoviruses have been found in many species of fishes and are large, enveloped DNA viruses. Of the four genera, only two, *Lymphocystivirus* and *Ranavirus*, cause disease in fishes [1]. Some iridoviruses, such as largemouth bass virus (LMBV) and gourami iridovirus, do not cause disease except in adverse environmental conditions. Several iridoviral diseases are listed by

---

* Corresponding author.
E-mail address: pettyd@mail.vetmed.ufl.edu (B.D. Petty).

doi:10.1016/j.cvex.2004.09.003   vetexotic.theclinics.com

the Office International des Epizooties as notifiable diseases, but they are limited to foodfish such as red sea bream and white sturgeon which are not kept commonly in home aquaria. Many iridoviral isolates from fish have low pathogenicity.

Lymphocystis

Lymphocystis disease, caused by an iridovirus, is the most common viral disease that is seen in pet fish. It affects many different fish species—freshwater and marine—and has worldwide distribution. The most commonly affected ornamental fishes are the glassfish (*Chanda ranga*) and the marine angels (Pomacanthidae). Other susceptible fishes include cichlids, butterflyfishes (Chaetodontidae), and sunfishes (Centrarchidae). Catfish and cyprinids are not susceptible.

Infection with either of two viruses, lymphocystis disease virus type 1 or type 2, will result in similar clinical signs [1]. These viruses are the only members of the genus *Lymphocystivirus* and are the largest of the iridoviruses; they range in size from 130 nm to 330 nm.

Both viruses infect the dermal fibroblasts and result in hypertrophy that gives rise to the lesions that are visible grossly. The skin of susceptible fish that are infected with either virus will have multiple focal to coalescing irregular gray or white masses (Fig. 1). Fins are affected commonly, although the lesions also may occur on the body. Infrequently, lymphocystis has been found to infect internal organs of some marine fishes [2].

Wet mounts of skin scrapings reveal the hypertrophied cells, as will histopathology of an infected site (Fig. 2). Lymphocystivirus is the only virus in fish that can be diagnosed confidently with wet mount preparations of skin scrapings and histopathology of affected areas.

It is uncommon for fish to die from this viral infection. Frequently, the lesions disappear after a period of time; however, if lesions are located near the mouth, feeding may be disrupted with subsequent starvation.

Fig. 1. A marine angelfish (*Centropyge fisheri*) displaying the typical gray masses of lymphocystis. (Courtesy of Stephen A. Smith, DVM, PhD, Blacksburg, VA.)

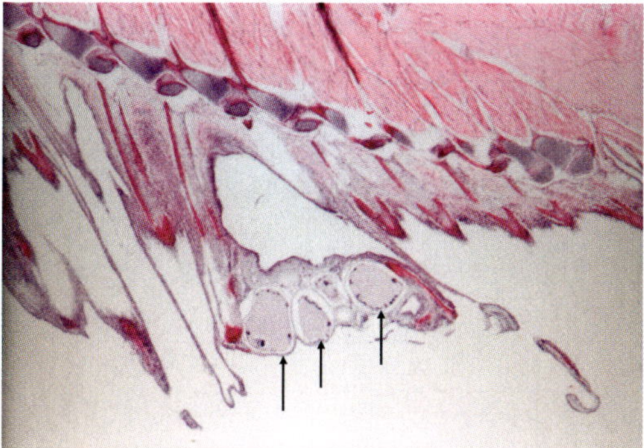

Fig. 2. Tissue section of juvenile clownfish (*Amphiprion* sp). Note the greatly hypertrophied cells (*arrows*) due to infection with lymphocystis virus (hematoxylin-eosin, original magnification ×40).

The virus is transmitted horizontally, when infected skin cells rupture and release infective particles into the water. The virus gains entry to a susceptible fish at skin and gills. Dense fish populations that are subject to regular handling (such as at a wholesale or retail facility) are at high-risk for lymphocystis transmission. Even in such situations, it is not unusual to observe infections on only certain species.

Experimentally, the incubation time that was required for lesions to develop in infected fish decreased as water temperature increased. Lesions developed in 5 to 12 days at 20°C to 25°C [3]. In wild marine fish populations, incidence is frequently the highest in spring and summer.

Neither of the lymphocystiviruses is grown easily in cell culture because of the long incubation period that is required. Cell lines that have been used successfully include bluegill fry (BF-2) and red drum dorsal fin cells. Electron microscopy of negative stains of the masses reveal viral particles (Fig. 3). Stress reduction via maintenance of good water quality conditions and use of gentle handling techniques may minimize impact of the virus and aid resolution of the lesions.

Viruses that affect cyprinids

Spring viremia of carp virus

Spring viremia of carp virus (SVCV) was officially recognized first by Fijan in 1971 as the causative agent of an acute hemorrhagic infection that primarily affects carp, although fish that exhibited typical signs of SVCV infection were first described in 1904 [4]. Infection with this virus has resulted in 30% to 70% mortality rates of cultured and wild carp in Europe. Although there have been no reports of infected fish in Asia, SVCV-positive fish were identified in a shipment of koi and goldfish that was sent to England from China in the late

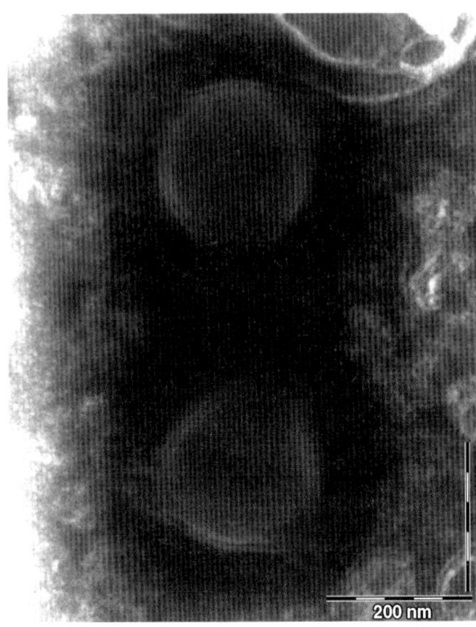

Fig. 3. Electron micrograph of lymphocystis viral particles from a negatively stained skin mass from a clownfish (*Amphiprion* sp) (negative stain phosphotungstic acid (PTA), original magnification ×50,000).

1990s [5]. Two separate cases of SVCV occurred in the United States in 2002. One case was identified as the cause of low mortality in koi carp at a koi production facility in North Carolina. The other case occurred in feral carp in Cedar Lake, in Wisconsin. In April 2004, a positive case was identified in a koi hobbyist pond in Washington. Another case was identified in July 2004 at a production facility in Missouri in koi that had been transported from Minnesota. Ongoing surveillance programs that are performed by the U.S. Fish and Wildlife Service and the U.S. Department of Agriculture (USDA) of wild and cultured susceptible fishes, respectively, have not revealed any SVCV-positive fish. Trade in ornamental koi of unknown SVCV status continues, so it is possible that more cases will occur in the United States. The USDA is considering the implementation of a health certification requirement for import of SVCV-susceptible species. This program would require that all susceptible fish that enter the United States originate from a certified SVCV-free facility/zone/country and that a certificate of veterinary inspection that attests to the negative SVCV status accompanies the fish.

The causative agent of spring viremia of carp is an RNA virus (formerly Rhabdovirus carpio) that is tentatively placed in the genus vesiculovirus in the Rhabdoviridae family. It is one of several rhabdoviruses that cause disease in fish, although SVCV is the only rhabdoviral disease of ornamental fish. At the time of this writing, spring viremia of carp is the only notifiable viral disease of ornamental fish. Any positive case in the United States must be reported to the state veterinarian. Susceptible fishes are koi/common carp (*Cyprinus carpio*), grass carp (*Ctenopharyngodon idella*), bighead carp

(*Hypophthalmichthys nobilis*), silver carp (*Aristichthys nobilis*), Crucian carp (*Carassius carassius*), and goldfish (*Carassius auratus*) [4].

Clinical signs are variable and can be compounded by concurrent infections (parasitic, bacterial, or fungal). Infected fish frequently exhibit lethargy, decreased respiration, loss of equilibrium, exophthalmia, pale gills, hemorrhages, ascites, protruding vent, and skin ulcers. Common internal findings include edema, inflammation, and pin-point hemorrhages (especially swim bladder). Multiple white foci on the atrium were present in SVCV-infected fish in Switzerland [6].

Perivasculitis in hepatic vessels and multi-focal necrosis in liver and pancreas are common findings on histopathology. Excretory and hematopoietic tissue of kidney also may be affected. Perivasculitis, desquamation of epithelium, and villar atrophy may be observed in the intestine.

Transmission primarily is horizontal, with infective particles found in feces, gill and skin mucus, urine, and exudate from skin blisters. Vertical transmission may occur, but is not common [7]. The virus can be transmitted by vectors, such as blood-sucking parasites, fish-eating birds, or fish-handling equipment.

The virus enters by way of the gills, which are the primary site of replication. Temperature is the most important factor in the determination of whether fish exhibit clinical signs. Disease occurs when the water temperature is 15°C to 20°C. If susceptible fish are exposed to SCVC and the water temperature is warmer than 20°C, they typically do not exhibit clinical signs.

The virus only can be isolated from infected fish when the water temperature is colder than 20°C. Therefore, it is important to determine if the water temperature is at or below 20°C when collecting suspect SVCV specimens.

Identification methods include cell culture of liver, spleen, and kidney, using epithelioma papillosum of carp (EPC) or fathead minnow (FHM) cell lines, immunofluorescent antibody technique, ELISA, virus neutralization, and PCR. On cell cultures, cytopathic effect (CPE) consists of widespread cell rounding followed by release of cells from plate surface [8]. Virus neutralization is the confirmatory test for SVCV at the time of this writing. Veterinarians who work with SVCV-susceptible fish should ascertain that the laboratory that is used for viral diagnostics is approved by the USDA to perform diagnostic tests for SVCV.

There is no treatment for SVCV; if it is diagnosed in a fish facility in the United States, all exposed fish, including fish species that are not susceptible to SVCV, must be depopulated. The virus can survive in the environment without a host for up to 42 days in mud, so disinfection and sanitation must be directed at eradicating it from the environment. SVCV is destroyed easily with disinfectants (eg, chlorine, formalin, sodium hydroxide) and pH extremes of less than 4 and greater than 10. Ozonation and ultraviolet irradiation are effective methods to disinfect potentially infected water sources.

Ideally, all susceptible species should come from an SVCV-free source. If this is not possible, they should be quarantined for a minimum of 30 days. Manipulation of water temperature to exacerbate clinical signs of disease may be useful during the quarantine period.

To qualify as an SVCV-free source, a facility must undergo SVCV testing. A random group of 150 susceptible fish is selected for testing, only when the water temperature is in the appropriate range for expression of clinical signs (10°C–18°C), by an accredited veterinarian. The tests must be conducted twice yearly. After 2 years of negative tests, the facility can be declared to be free of SVCV. To maintain that designation, the facility must be tested annually.

The potential impact of this disease is significant. If a facility has a positive fish and does not have good animal movement records, all fish at that facility may be depopulated. The loss of valuable fish, potential loss of sales, labor costs for disinfection, and other costs can be a devastating blow to a business.

Koi herpesvirus

The first case of koi herpesvirus (KHV) was diagnosed as the cause of massive mortality of ornamental koi in the United States in 1998 [9]. Since that time, Israel, Europe, and Asia also have reported cases in ornamental koi. In 2003, massive mortalities that were due to KHV occurred in wild common carp in Japan.

The only susceptible species to KHV infection is common carp/koi carp (*Cyprinus carpio*). Clinical signs of KHV infection are nonspecific, but include patchy white areas on the gills that resemble columnaris infections and production of excess mucus on the body. KHV-infected fish also may have secondary parasitic and bacterial infections. Necrosis of internal organs is a common finding on internal examination. Infected fish may swim near the surface of the water and act lethargic.

The agent of KHV, a DNA virus, is classified as a member of the Herpesviridae family (Fig. 4) [10]. It is transmitted horizontally by way of contact with infected fish or water that is contaminated with the virus. The virus is infective in water for up to 4 hours [11]. It is killed easily with standard disinfectants.

If an infected fish is introduced into a population of susceptible fish, clinical signs of disease may be exhibited 14 days later. Mortality rates of 80% to 100% may occur in susceptible fish when the water temperature is 18°C to 27°C. These mortality rates may be decreased or halted by increasing the water temperature to 30°C; however, this is not recommended because surviving fish may become a carrier population. Another herpesvirus of fish, channel catfish virus (CCV), serves as a good example. Catfish that survive CCV outbreaks were proven to be a covert source of virus to susceptible catfish [3]. Further research is needed to determine if KHV is similar; until proven it is best to assume that a carrier state can exist.

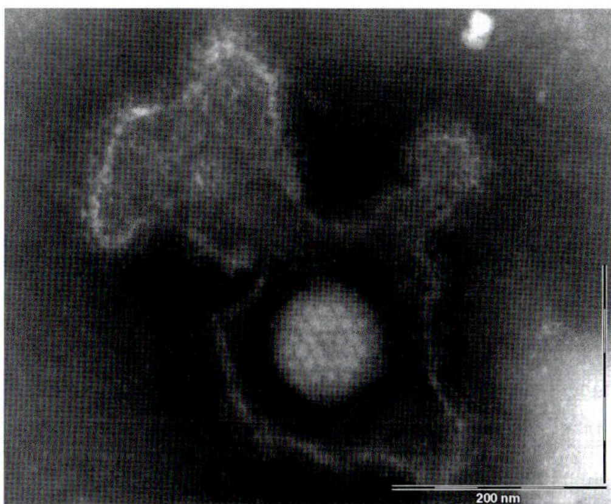

Fig. 4. Electron micrograph of koi herpesvirus from infected koi fin–1 cell line (negative stain PTA, original magnification ×140,000).

To diagnose KHV, gill, spleen, and kidney samples should be submitted for PCR and cell culture. KHV will grow on the koi fin cell line (KF-1) and carp brain cell line [12]. Typical CPE consists of vacuole formation and eventual detachment of the cells (Fig. 5). The PCR method is a more sensitive and rapid diagnostic method than virus isolation.

Histopathologic lesions that are caused by KHV may be extensive. Fusion of secondary lamellae and hyperplasia, hypertrophy, and necrosis of epithelial cells have been observed in the gills. Focal necrosis and intranuclear inclusions occurred in the splenic parenchyma and in the hematopoietic cells of the kidney.

New koi should be quarantined for at least 30 days and the water temperature should be maintained between 18°C and 27°C during the quarantine period to promote expression of KHV infection, if present.

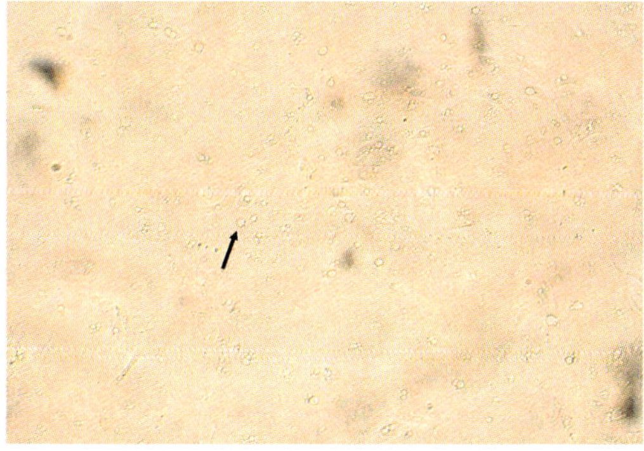

Fig. 5. KF-1 cell line displaying typical signs of KHV infection. Note the vacuole formation (*arrow*) (original magnification ×100).

Methods for determining previous exposure to KHV, such as in situ hybridization and serology for anti-KHV antibodies, are being developed (A. Goodwin, PhD, Pine Bluff AR, personal communication, September 2004).

Cyprinid herpesvirus

Cyprinid herpesvirus (CHV-1), which also is known as carp pox, fish pox, or epithelioma papillosum, is a benign, hyperplastic condition that is limited to the epidermis and primarily affects adult common carp/koi carp (*Cyprinus carpio*). The disease is common to all areas of carp culture.

In contrast to lymphocystis lesions which appear nodular and rough, the lesions that are caused by carp pox are smooth and waxlike in appearance. The lesions are opaque and raised and may cover most of the body and fins.

Mortalities that are due to CHV-1 are rare in older fish, but mortality rates may be high in juvenile fish. In older fish, the lesions may appear seasonally, usually when the water temperature is colder than 15°C. Lesions regress as the water temperature is warmed to 20°C. After lesion regression, the genome of the virus has been found in spinal nerves, cranial nerve ganglia, and subcutaneous tissue.

The virus is transmitted horizontally by way of cohabitation as a result of release of viral particles from ruptured skin cells. Although the virus has been detected in the eggs of infected female carp, evidence of disease in fry from infected eggs has not been observed. CHV-1 has been transmitted experimentally by way of intracoelomic and intramuscular injections.

The lesions that are caused by CHV-1 can be differentiated from those that are caused by lymphocystis virus by light microscopy. CHV-1 causes hyperplasia of the epidermal cells, whereas lymphocystis virus causes hypertrophy. In juvenile carp, necrosis of the kidney, liver, and intestine may be observed.

CHV-1 is caused by *Herpesvirus cyprini*, which readily grows on EPC or FHM cells. CPE includes vacuole formation, rounding of the cells, and focal pyknosis. Clinical signs of infection can be managed by manipulating the water temperature to reduce the lesions.

Grass carp hemorrhagic virus

Grass carp hemorrhagic virus (GCHV) is primarily a disease of grass carp in China. Goodwin's [13] comparative analysis of GCHV and golden shiner reovirus revealed that the viruses are one and the same. Because veterinarians may be asked to assist with diagnostic testing of golden shiners (*Notemigonus crysoleucas*), which are used commonly as baitfish, information on GCHV and its effect on golden shiners are included here.

An aquareovirus, GCHV was first isolated from golden shiners in the United States in 1977 [3]. It is an icosahedral RNA virus (Fig. 6). It has been associated with disease in golden shiners only when the water temperature

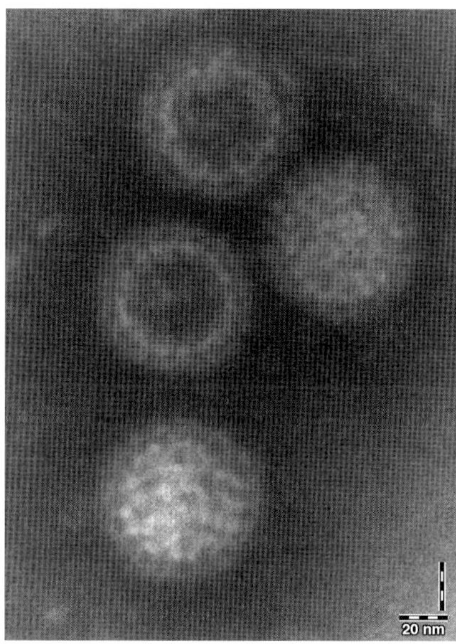

Fig. 6. Electron micrograph of golden shiner reovirus particles (negative stain PTA, original magnification ×140,000).

exceeds 25°C. A high stocking rate may facilitate progression of disease. Losses may be as high as 50% if stocking density is high.

The virus is transmitted horizontally, with an incubation period of 1 to 2 weeks. It is infective in water for 15 days at 4°C to 30°C [7]. Infected fish may act lethargic and swim near the surface. Occasionally, petechial hemorrhage in the skin may be observed.

Kidney and spleen are the optimum organs for virus isolation. GCHV grows on FHM at 28°C to 30°C. The severity of the disease can be reduced by decreasing the stocking rate and decreasing water temperature, if possible.

Viruses that affect cichlids

Iridovirus-associated disease in angelfish

In the late 1980s and early 1990s, mortalities in freshwater angelfish (*Pterophyllum scalare*) soared. Hobbyists, retail stores, and producers experienced higher losses than normal. Rumors abounded that an unknown agent was responsible for "angelfish plague" or "angelfish AIDS." Researchers at the University of Florida analyzed multiple cases of mortality in angelfish during that period. Water quality problems were the most common source of disease in the fish that were submitted for diagnostic testing. Ectoparasites were the next most common source of disease. Iridoviruses were seen in a few cases; however, none could be isolated in cell culture (Fig. 7) [14].

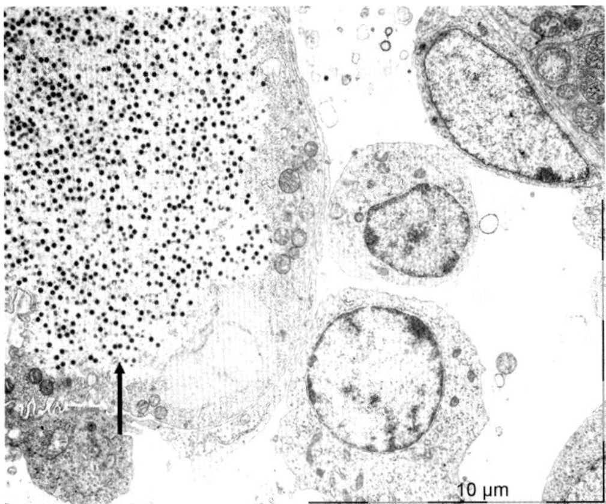

Fig. 7. A hallmark of iridovirus disease is the presence of swollen, virus infected cells. In lymphocystis, the ballooned cells occur in the skin and the result usually is not fatal. In this electron micrograph (negative stain), an iridovirus infected cell (*arrow*) in an angelfish kidney is many times larger than its uninfected partners. When such abnormally large cells occur internally, the result frequently is fatal (epoxy section, lead citrate, and uranyl acetate, original magnification ×3200).

Schuh and Shirley [15] reported a case of an iridovirus in one juvenile angelfish. Extensive necrosis and hemorrhage of the hematopoietic tissue in the kidney and spleen and pancreatic necrosis were observed on histology. Hypertrophied cells, with an occasional inclusion body, were seen in association with the necrosis. On examination by transmission electron microscopy, the hypertrophied cells contained intracytoplasmic iridovirus particles.

In contrast, in 2003, a commercial producer of freshwater angelfish began to experience increased mortality rates that were limited to two vats of the facility. Fish were submitted to a laboratory for examination. The affected fish had enlarged abdomens and abnormal body posture (head angled up). Moderate numbers of flagellated protozoans were observed in the distal intestine, but no other parasite was observed. Bacterial cultures of spleen, kidney, and brain were negative. Electron microscopic examination of spleen and kidney revealed iridovirus particles; however, there was no associated pathology on light microscopy. In this case, the virus seems to have been an incidental finding. The producer elected to depopulate the fish in the facility, disinfected the systems, and purchased new fish. No similar losses have been reported (unpublished data).

Herpesvirus-associated disease in angelfish

In addition to the finding of iridovirus in angelfish that experienced mortality during the 1990s, a herpesvirus also was associated with increased mortality—mostly in juveniles—during movement from ponds to tanks.

Excess mucus on the skin was a common finding in the diseased fish. Bacteria and parasites were ruled out as pathogens and there were no detectable water quality problems. Several cases with similar histories all had confirmed herpesvirus infections; this strongly implicated the virus as the cause of death. Attempts to isolate the virus in cell culture were unsuccessful, although viral particles were observed on electron microscopy of the skin (Fig. 8).

Retrovirus-associated disease in angelfish

Frequently, retroviruses cause neoplasia in fish and other vertebrates [1]. Enveloped RNA viruses, they are linked to several neoplastic diseases in fish—walleye dermal sarcoma virus, damselfish neurofibromatosis virus, salmon leukemia virus, and angelfish retrovirus. Occasionally, angelfish (*P scalare*) may present with large nodular masses around the mouth. The tumors are unsightly, but usually do not result in mortality; however the ability to take up food particles may be impaired and subsequent weight loss may occur.

Francis-Floyd et al [16] reported that histologic examination of the tumors revealed dense fibrovascular tissue that was covered by thickened stratified squamous epithelium. Retrovirus-like particles were seen in the cytoplasm of neoplastic stromal cells on ultrastructural examination.

Attempts to infect healthy angelfish by injection with cell-free ultra-filtrates or topical application of the ultrafiltrate to abraded lips were not successful. The fibromas can be removed surgically and there usually is no recurrence [17].

Iridovirus-associated disease in oscars

After a heater malfunctioned in one system at an oscar (*Astronotus ocellatus*) production facility in Florida, water temperature increased from

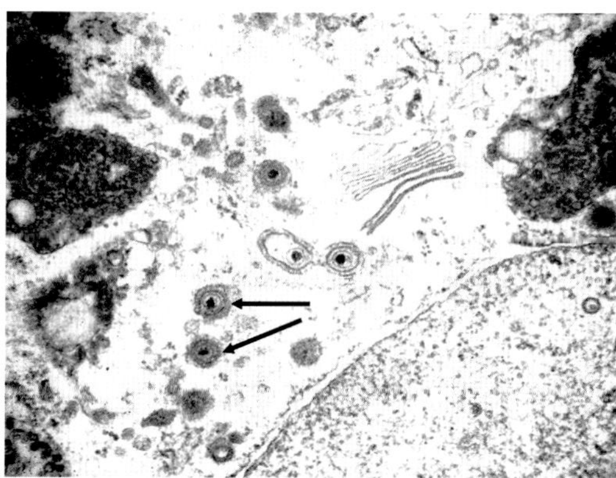

Fig. 8. Electron micrograph of herpesviral particles (*arrows*) in skin of angelfish. (negative stain; epoxy section, lead citrate, and uranyl acetate; original magnification ×20,000).

82°F to 94°F (28°C to 37°C) during a 2- to 3- day period [18]. The heater was repaired and the water temperature returned to normal. Several days after the temperature increase, fish in the system began to die; mortality rates ranged from 20% to 100% in the different tanks of the system.

Clinical signs of the fish that were submitted to a diagnostic laboratory included exophthalmia, darkening of the skin, and skin ulcerations. Behavioral abnormalities included hanging at the surface, lying in lateral recumbency at the bottom, and spinning.

Although numerous motile bacteria were observed on skin of the fish with skin ulcerations, no significant pathogen was identified on wet mount examination of external and internal organs or bacterial culture of posterior kidney and brain. Acute, diffuse, severe coagulative necrosis with intra-cytoplasmic inclusions in the kidney was the primary tissue lesion that was observed in fixed specimens. In addition, intracytoplasmic inclusions were observed in other tissues, including endothelial cells of the blood vessels of the gills and in atrial tissue, and interstitial cells in the esophagus, skeletal muscle, and bone. Ultrastructural examination of the affected tissues revealed iridoviral particles. Viral isolation was not attempted. The producer de-populated and disinfected the entire facility and purchased new broodstock. No similar problem has been reported since.

Viruses that affect anabantoids

Gourami iridovirus

In the early 1990s, several gourami producers in Florida began to experience mortality rates that varied from 10% to as high as 70% in the summer months. The three spot gourami (*Trichogaster trichopterus*), a popular gourami in the pet trade, was the most common species that was affected. Iridoviruses also have been observed in dwarf gouramis (*Colisa lalia*), kissing gouramis (*Helostoma temminckii*), and pearl gouramis (*T leeri*), but it is unknown if those viruses are the same as the one in three spot gouramis.

Affected fish displayed patchy hyperpigmentation and were lethargic (Fig. 9). Some fish also had ascites. On internal examination, the spleens of affected fish were enlarged and a clear, amber fluid was present in the coelom. No parasitic or bacterial pathogen was isolated.

On electron microscopic examination of internal tissues, iridovirus particles were observed in the spleen, intestine, and anterior kidney (Fig. 10). Viral isolation was attempted by culturing spleen and intestine on FHM and tilapia heart (TH) cells. CPE occurred in the TH cells 3 to 5 days postinoculation. Electron microscopic examination of the TH cultures revealed iridovirus particles [19].

On histopathology, the primary lesion was diffuse necrosis of the splenic parenchyma. Intracytoplasmic and intranuclear inclusions were present in

Fig. 9. Photograph of a three spot gourami with iridovirus infection that displays hyperpigmentation.

splenic cells; intracytoplasmic inclusions also were observed in the hematopoietic cells of the anterior kidney.

Lynch [20] demonstrated that the iridovirus could be transmitted horizontally. Gourami iridovirus–negative, noninjected fish that cohabited with fish that were injected with gourami iridovirus began to die several days after the infected fish began to exhibit clinical signs of infection. Warm water temperatures ($\geq 32^\circ$C) result in increased mortality, but it is unknown whether this effect is due to suppressed immune system or enhanced viral replication. When Lynch injected healthy gouramis with gourami iridovirus and maintained them at two different water temperatures (23°C and 32°C), mortalities occurred at 32°C, but not at 23°C. When the water temperature was increased from 23°C to 32°C, the injected fish began to die.

There is no treatment, but mortality can be reduced or avoided by not exposing fish to water temperatures that are 32°C or warmer. In production ponds in which water temperature is difficult to control, fish should not be handled if the water temperature is at least 32°C. Use of shade cloth and running water into the ponds can help to reduce water temperature.

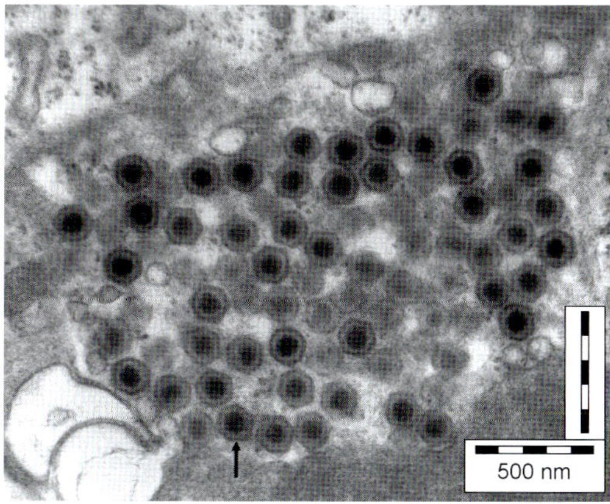

Fig. 10. Electron micrograph of gourami iridovirus (*arrow*) from spleen (epoxy section, lead citrate, and uranyl acetate, original magnification ×32,500).

Viruses that affect catfishes

Pleco herpesvirus

Herpesvirus like particles have been seen in two species of armored suckermouth catfishes, the blue-eyed pleco (*Panaque suttoni*) and the common pleco (unidentified loricariid).

A veterinarian who worked with an ornamental production and wholesale facility observed that blue-eyed plecos (an imported fish) occasionally arrived with pale, blotchy lesions on the skin [21]. Over a period of several days, the lesions coalesced and appeared to be ulcerative and the affected fish died within days of the appearance of the lesions (Fig. 11). Herpesvirus like particles were observed by electron microscopic examination of the skin lesions. This species is not imported commonly and no new cases have been seen.

The common pleco is a popular fish that is kept by hobbyists for algae control and scavenging leftover food. The eggs are collected from creeks, canals, and rivers and sold to farmers who hatch and grow the plecos to a sellable size. Some farmers noticed the development of blotchy lesions on the skin of the plecos when the water temperature was lower than the optimum temperature for the species. Affected plecos were submitted for diagnostic testing; herpesvirus particles were seen on negative stains of the skin lesions. Attempts to isolate the virus in cell culture were not successful.

Viruses that affect centrarchids

Largemouth bass iridovirus

Although largemouth bass (*Micropterus salmoides*) usually are not considered a pet fish, they are kept frequently as display animals in bait shops and outdoor equipment stores. Largemouth bass virus (LMBV) was

Fig. 11. Blue-eyed plecos infected with a herpesvirus. Note the vesicles and coalesced lesions. (Courtesy of Roy P.E. Yanong, VMD, Ruskin, FL.)

identified first in healthy, as well as diseased, largemouth bass during an investigation of a fish kill in a Florida lake in 1991. No pathology was associated with the virus in any of the fish [22]. In 1995, LMBV was isolated from two moribund largemouth bass from a large fish kill in the Santee-Cooper Reservoir in South Carolina [23]. Although LMBV has been implicated in sporadic fish kills during the summer months around the southeast United States, no fish kills have been attributed to the presence of LMBV in Florida where approximately 30% of the lakes that were sampled had largemouth bass that were positive for LMBV. Recent studies suggest a link between high water temperature and expression of clinical signs of LMBV infection [24].

In addition to largemouth bass, susceptibility to LMBV is reported in Florida largemouth bass (*M salmoides floridanus*), a subspecies of large-mouth bass. Other centrarchids, such as bluegill (*Lepomis macrochirus*), are reported to be susceptible to LMBV, but do not show signs of illness. The virus is spread by way of horizontal transmission; Woodland et al [25] had limited success when they attempted to transmit the virus orally. There is no evidence that vertical transmission occurs.

LMBV has been blamed for fish kills of trophy-sized largemouth bass; however, injection of LMBV into adult largemouth bass did not result in disease, but 100% of juvenile largemouth bass died following injection [23,26]. Reports indicate that the swim bladder of infected fish may be filled with an exudate or may be inflamed and overfilled [23].

LMBV easily grows in FHM (Fig. 12) or BF-2 cells. CPE consists of pyknotic rounded cells that eventually detach. Serology and PCR methods also are available. More research is required to demonstrate that LMBV is a causative agent in largemouth bass kills.

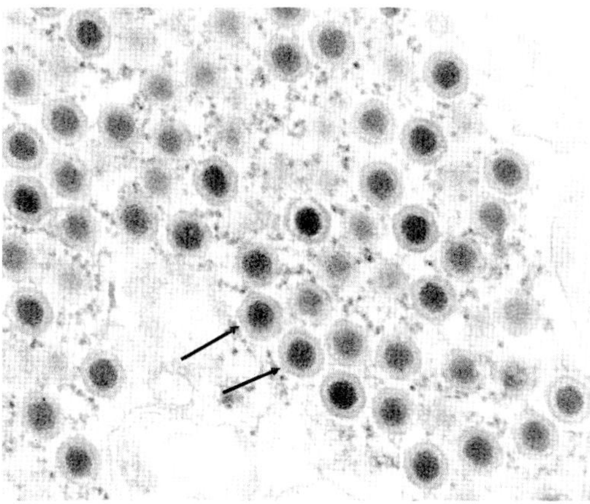

Fig. 12. Iridovirus particles (*arrow*) in FHM cell culture from the first isolation of LMBV. Virus was grown from fish tissues, but virions were never found in the original tissue which suggested that they were present in low numbers (epoxy section, lead citrate, and uranyl acetate, original magnification ×32,500).

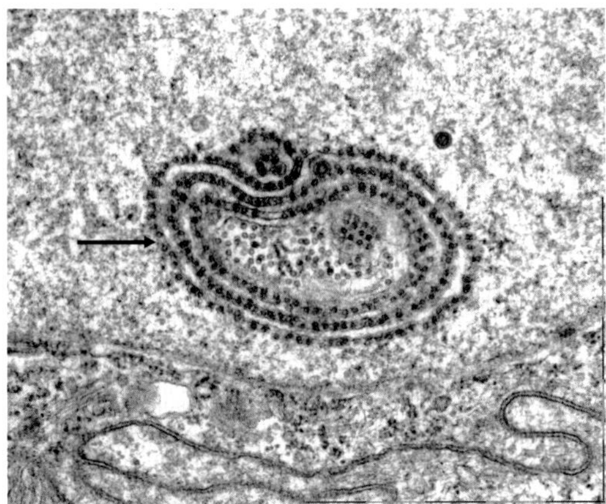

Fig. 13. Whirling behavior in rainbowfish was associated with large numbers of picornalike virus particles (*arrow*) in the brain. Electron micrograph, negative stain (epoxy section, lead citrate, and uranyl acetate, original magnification ×20,000).

Other viruses

Picornavirus in rainbowfish

Turquoise rainbowfishes (*Melanotaenia lacustris*) that exhibited whirling symptoms and obvious impairment of the central nervous system were submitted by a producer for diagnostic evaluation. All were negative for bacterial and parasitic pathogens. Electron microscopy revealed numerous picornavirus particles in the brain (Fig. 13). Attempts to grow the virus in cell culture were unsuccessful. No similar cases have been reported since. Picornaviruses have been observed as incidental findings in various freshwater fishes. The significance of their presence is unknown.

Summary

Investigation of a fish kill should begin with water quality analysis and use of standard necropsy techniques to rule out the role of parasites or bacteria. The presence of a virus does not mean necessarily that it is a pathogen. Another point to consider is the potential influence of water temperature on pathogenicity; this may be an important factor in developing disease.

A list of laboratories that is approved by the USDA to conduct diagnostic testing of aquaculture species can be found at: http://www.aphis.usda.gov/vs/nvsl/labcertification/aquaapplab.htm.

References

[1] Chinchar VG. Ecology of viruses of cold-blooded vertebrates. In: Hurt CJ, editor. Viral ecology. London: Academic Press; 2000. p. 413–45.

[2] Colorni A, Diamant A. Splenic and cardiac lymphocystis in the red drum, *Sciaenops ocellatus* (L.). J Fish Dis 1995;18:467–71.

[3] Plumb JA. Health maintenance and principal microbial diseases of cultured fish. Ames (IA): Iowa State University Press; 1999.

[4] Wolf K. Fish viruses and fish viral diseases. Ithaca (NY): Cornell University Press; 1988.

[5] Hoole D, Bucke D, Burgess P, et al. Diseases of carp and other cyprinid fishes. Oxford (UK): Fishing News Books; 2001.

[6] Bernet D, Wahli T. First report of SVC infection in koi carp in Switzerland. Bull Eur Assoc Fish Pathol 2002;22(3):225–9.

[7] Fijan N. Spring viraemia of carp and other viral diseases and agents of warm-water fish. In: Woo PTK, Bruno DW, editors. Fish diseases and disorders: viral, bacterial, and fungal infections, vol. 3. New York: Cab International; 1999. p. 177–246.

[8] Goodwin AE. First report of spring viremia of carp (SVCV) in North America. J Aquat Anim Health 2002;14(3):161–4.

[9] Hedrick RP, Gilad O, Yun S, et al. A herpesvirus associated with mass mortality of juvenile and adult koi, a strain of common carp. J Aquat Anim Health 2000;12(1):44–57.

[10] Waltzek TB, Kelley GO, Yun SC, et al. Relationships of koi herpesvirus (KHV) to herpes-like viruses of fish and amphibians. Presented at the 35th Annual Conference, International Association for Aquatic Animal Medicine. Galveston, Texas, April 4–9, 2004.

[11] Perelberg A, Smirnov M, Hutoran M, et al. Epidemiological description of a new viral disease afflicting cultured *Cyprinus carpio* in Israel. The Israeli Journal of Aquaculture 2003; 55(1):5–12.

[12] Gray WL, Mullis L, LaPatra SE, et al. Detection of koi herpesvirus DNA in tissues of infected fish. J Fish Dis 2002;25:171–8.

[13] Goodwin AE. Molecular, physical and clinical evidence that golden shiner virus (GSV) and grass carp reovirus (GCR) are variants of the same virus. J Aquat Anim Health 2003;15(4): 257–63.

[14] Francis-Floyd R, Fraser W, Yanong R. Iridoviruses in Florida fisheries. Presented at the First Joint Meeting, International Association of Aquatic Animal Medicine and American Association of Zoo Veterinarians, New Orleans, September 17–21, 2000.

[15] Schuh JCL, Shirley IG. Viral hematopoietic necrosis in an angelfish (*Pterophyllum scalare*). Jour Zoo Wild Med 1990;21(1):95–8.

[16] Francis-Floyd R, Bolon B, Fraser W. Lip fibromas associated with retrovirus-like particles in angel fish. J Am Vet Med Assoc 1993;202(3):427–9.

[17] Harms CA, Lewbart GA. Surgery in fish. Veterinary Clin North Am Exot Anim Pract 2000; 3(3):759–74.

[18] Yanong RPE. Iridoviral-associated disease in oscars (*Astronotus ocellatus*). Presented at the 34th Annual Conference, International Association of Aquatic Animal Medicine. Waikoloa, Hawaii, May 10–14, 2003.

[19] Fraser WA, Keefe TJ, Bolon B. Isolation of an iridovirus from farm-raised gouramis. J Vet Diagn Invest 1993;5:205–53.

[20] Lynch MJ. Preliminary investigation of gourami iridovirus infections [master's thesis]. Gainesville (FL): University of Florida; 1998.

[21] Yanong RPE. Possible herpesvirus-associated disease in the blue-eyed plecostomus, *Panaque suttoni*. Presented at the 26th Annual Conference, International Association of Aquatic Animal Medicine. Mystic, Connecticut, May 6–10, 1995.

[22] Grizzle JM, Altinok I, Fraser WA, et al. First isolation of largemouth bass virus. Dis Aquat Organ 2002;50:233–5.

[23] Plumb JA, Grizzle JM, Young HE, et al. An iridovirus isolated from wild largemouth bass. J Aquat Anim Health 1996;8(4):265–70.

[24] Grant EC, Philipp DP, Inendino KR, et al. Effects of temperature on the susceptibility of largemouth bass to largemouth bass virus. J Aquat Anim Health 2003;15(3): 215–21.

[25] Woodland JE, Brunner CJ, Noyes AD, et al. Experimental oral transmission of largemouth bass virus. J Fish Dis 2002;25:669–72.
[26] Plumb JA, Zilberg D. The lethal dose of largemouth bass virus in juvenile largemouth bass and the comparative susceptibility of striped bass. J Aquat Anim Health 1999;11: 246–52.

ELSEVIER
SAUNDERS

Vet Clin Exot Anim 8 (2005) 85–105

VETERINARY
CLINICS
Exotic Animal Practice

Viral diseases of companion birds

Cheryl B. Greenacre, DVM, DABVP-Avian

*Department of Small Animal Clinical Sciences, College of Veterinary Medicine,
University of Tennessee, C247 2407 River Drive, Knoxville, TN 374996, USA*

The term "companion bird" in this article includes parrots, pigeons, canaries, and finches. Many viruses are known to cause disease in companion birds and have been well-studied. Other avian viruses have not been well-studied and some are not yet classified; probably others exist that have not been discovered because not all known viral families have been documented in birds. The following viral families are discussed alphabetically: Adenoviridae, Circoviridae, Coronaviridae, Flaviviridae, Herpesviridae, Orthomyxoviridae, Papovaviridae, Poxviridae, Reoviridae, Retroviridae, Rhabdoviridae, Togaviridae, and a virus from an unclassified virus family, proventricular dilatation disease. When appropriate, each virus family includes a description of the virus, the species affected, the transmission route and recommended type of disinfection, clinical signs, diagnosis, treatment, prognosis, and prevention modalities available.

Adenoviridae

Adenoviruses are nonenveloped viruses that are known to infect vertebrates. Avian viruses in the Adenoviridae family are in the genus Aviadenovirinae and include duck hepatitis, quail bronchitis, pigeon adenovirus, turkey viral hepatitis, goose adenovirus, marble spleen disease of pheasants, egg drop syndrome, turkey hemorrhagic enteritis, and the newly described hydropericardium syndrome in poultry [1]. Viruses in the Aviadenovirinae genus are separated into strains based on virus neutralization tests. The following description focuses on pigeon adenovirus, which is a tentative species within the genus Aviadenovirinae, and an adenovirus of psittacine birds that is not assigned to a genus according to the International Committee on Taxonomy of Viruses [2]. Pigeons are susceptible to some serotypes that infect chickens and some serotypes that are pigeon specific.

E-mail address: cgreenac@utk.edu

Recently, fowl adenovirus serotype 4, strain KR5 was isolated from African gray parrots and cape parrots that had clinical disease [3].

Transmission of adenoviruses can occur horizontally, through the fecal oral route, and vertically. Vertical transmission has been demonstrated in chickens. Latent infections with periodic shedding are believed to occur.

Generally, adenoviruses cause the most severe lesions in birds that are less than 1 month of age. Most often, aviadenoviruses are considered to be opportunistic pathogens and usually are involved with concomitant infections or immunosuppression.

Pigeons

Pigeons from hatching to 5 years of age are susceptible to pigeon adenovirus; clinical signs are seen most commonly in birds that are 2 to 4 months of age. Clinical signs include depression, anorexia, dyspnea, a crouched stance, polydipsia, polyuria, and a slimy green diarrhea [4]. Young birds can die within 48 hours of displaying clinical signs. Mortality ranges from 0% to 60% and is greatest at 3 to 4 days postinfection [5,6].

Psittacine birds

Clinical signs in psittacine birds include depression, anorexia, and cloacal hemorrhage, followed by hepatitis, enteritis, pancreatitis, encephalitis, splenitis, conjunctivitis, and death [4]. Because birds that had no clinical signs exhibited intranuclear inclusion bodies that were consistent with adenovirus, it is suggested that asymptomatic infections can occur in psittacine birds. Usually budgerigars, lovebirds, cockatiels, Moluccan cockatoos, and red-rumped parakeets exhibit the typical clinical signs that were described above; however, other species can be found dead, including Amazon parrots, Patagonian cunures, Eastern rosellas, hyacinth macaws, lesser sulfur-crested cockatoos, and budgerigars. There is one report of an entire shipment of 59 eclectus that were shipped to a zoologic park dying of adenovirus [7]. This outbreak was associated with severe weight loss and liver disease that consisted of friable livers with whitish surface mottling and subcapsular hemorrhage.

Adenoviruses replicate in the nucleus and form basophilic intranuclear inclusion bodies that most often are recognized in hepatocytes and enterocytes [3]. Many different tests are available for poultry-associated aviadenoviruses, including ELISA, agar gel immunodiffusion (AGID), fluorescent antibody (FA), virus neutralization (VN), and hemagglutination inhibition (III); however, the best test that is commercially available for psittacine birds is the DNA in situ hybridization (this is offered at the University of Georgia's Infectious Disease Laboratory, Athens, Georgia) [8,9]. A polymerase chain reaction (PCR) test has been developed to detect pigeon adenovirus [10]. Treatment consists of supportive care.

Because adenoviruses are nonenveloped viruses, they remain infectious for a long time in the environment and are difficult to destroy with disinfectants. Exposure to formalin, aldehydes, or iodophores for longer than 1 hour is recommended to destroy adenoviruses. Free-ranging pigeons and waterfowl may acts as reservoirs to birds that are housed outdoors. There is no vaccine available for psittacine birds.

Circoviridae

In 1989, this new family of pathogenic animal viruses was described, including circoviruses that were isolated from cockatoos—called psittacine beak and feather disease (PBFD)—chicken anemia virus, and porcine circovirus [11]. These nonenveloped viruses are one of the smallest yet described (14–16 nm). Later, pigeon circovirus and goose circovirus were described; both are antigenically different from PBFD [12].

Psittacine beak and feather disease

The PBFD virus most commonly causes clinical signs in captive and free-ranging Old World (Australian and African) psittacine birds, such as cockatoos, lovebirds, African gray parrots, and cockatiels. Less commonly, it causes clinical signs in New World (Americas) psittacine birds, such as Amazon species, macaws, and conures. A few cases of clinical PBFD have been documented in New World species, including scarlet macaws, red-lored and blue-fronted Amazon parrots, and Janday conures [4,13]. The PBFD virus is endemic in many free-ranging flocks of psittacine birds in Australia.

PBFD virus is shed in feces, feather dander, and various excretions and secretions. Asymptomatic birds can shed the virus for years before exhibiting any clinical signs. Vertical transmission probably occurs as evidenced by the development of PBFD in artificially-incubated chicks from infected hens. Chicken anemia virus (CAV) was proven to be transmitted vertically to the eggs from hens that were infected during the laying period [4]. Because the virus is nonenveloped, it is stable, can survive years in the environment, and is resistant to destruction by common disinfectants. A cell culture–derived CAV was killed by 1% iodine, sodium hypochlorite, 0.4% B-propriolactone, 1% gluteraldehyde, or 80°C for 1 hour.

Peracute, acute, and chronic forms of PBFD occur in parrots. Generally, the progression of the disease is dictated by the age of the bird when clinical signs first appear. Younger birds have a faster progression of the disease.

Peracute PBFD is common in cockatoo and African gray parrot neonates and is associated with sepsis, pneumonia, enteritis, weight loss, and death. Some present with severe leukopenia and liver necrosis [14]. Histologic changes that are observed with peracute disease include feather follicle cell edema and bursal necrosis.

Acute PBFD affects chicks that are as young as 28 to 32 days of age. Clinical signs include depression and rapidly progressing feather dystrophy (necrosis, hemorrhage, and premature shedding of newly-developing feathers). In 1 week, 80% to 100% of feathers can be affected [4]. A nonregenerative anemia can be present, with a packed cell volume (PCV) of 14% to 25%.

Chronic PBFD is characterized by symmetric, slowly-progressive dystrophy of developing feathers that worsens with each successive molt. These are birds that survive the acute phase and then go on to have a prolonged incubation period that may involve years before clinical signs appear. The feather dystrophy includes retained feather sheaths, hemorrhage within the pulp, curled feathers, and circumferential constrictions of the feather shaft. Usually, the down and contour feathers are affected first, whereas the primary feathers are affected later. Birds can go on to develop complete alopecia, and sometimes, beak abnormalities that consist of progressive elongation of the beak and necrosis of the palate rostrally. These birds often are immunosuppressed and die of secondary bacterial or fungal infections.

Intracytoplasmic inclusion bodies are observed in the bone marrow, thymus and bursa. The DNA probe tests that are available at the University of Georgia Infectious Disease Laboratory can distinguish between PBFD and Psittacine Circovirus 2 (PsCV-2) of lories [15]. The PBFD DNA probe tests are performed on whole blood and detect viral DNA; a positive means that there was PBFD viral DNA in the blood. In a bird that has no clinical signs, it is recommended to retest the bird in 90 days to determine if the viral DNA is still present. If it is still present then the bird is infected, but if not, then the bird may have been infected transiently and overcame the infection. Any bird that displays feather abnormalities should have a feather follicle biopsy and DNA in situ hybridization performed in addition to the DNA probe blood test because some clinical birds are so viremic that they will have a negative blood test. This also can occur if a bird is extremely leukopenic [4]. A DNA probe test can be used to detect viral DNA on a swab of the environment to assist in determining the effectiveness of disinfection efforts. Treatment consists of supportive care and antimicrobials for secondary infections.

After clinical signs develop the disease is always fatal. Some exposed birds will mount an immune response and recover and never show clinical signs. These birds do not shed and are considered to be naturally vaccinated. In the United States, prevention through testing and isolation is the best way to control this disease. A vaccine is being developed for use in the United States. In Australia, an inactivated vaccine is available for use in birds, but it is not recommended for use in captive psittacine birds in the United States because of potential loss. In the United States it would be better to wait for a safe subunit or cell culture–derived vaccine and continue to test and isolate [4].

Psittacine beak and feather disease–2 and other strains

A PBFD variant (PsCV-2) has been described in lories; it is not as pathogenic as the originally-described PBFD [15]. Recently, a novel circovirus was described in mulard ducks that is associated with feather dystrophy [16]. Also, recent studies of PBFD viruses that were isolated from free-ranging birds in southern Africa demonstrated several unique genotypes that diverged substantially from PBFD viruses that were isolated from Australia and America [17].

Lories that had PsCV-2 had clinical feather lesions that were similar to PBFD but they had less severe clinical signs and, most importantly, they recovered. This is similar to what occurs with porcine circovirus 1 and 2 in pigs. One is highly pathogenic and the other is of little or no pathogenicity; therefore, it is imperative to distinguish between the two. Bird species other than lories may be infected with PsCV-2.

Pigeon circovirus

Pigeon circovirus is common in Europe. Clinically affected pigeons exhibit lethargy, anorexia, diarrhea, and poor weight gain. The highest mortality is seen in 7- to 8-week-old squabs. An outbreak occurred in pigeons that were 6 weeks to 12 months of age and many had concomitant bacterial, fungal, or parasitic infections. Intracytoplasmic inclusion bodies were observed in spleen, gut-associated lymphatic tissue, and cloacal bursa [12,18]. A DNA in situ hybridization and PCR tests have been developed to detect pigeon circovirus in tissue samples [19,20].

Coronaviridae

Parrots

In 1978, there was a report of a new coronavirus-like agent that was isolated from parrots. In 1982, these same investigators published a retraction of that claim because the samples were found later to be contaminated with chlamydial organisms [21,22]. Whenever the chlamydial organisms were filtered out of subsequent attempts to infect birds, the birds did not show any clinical signs of disease. Therefore, the investigators were unable to prove whether the clinical signs that were observed were due to the coronavirus-like agent that originally was observed on electron microscopy (EM) or to chlamydophila [22]. No further reports have been published.

Pigeons

There is one report of infectious bronchitis virus in a flock of racing pigeons in Australia that was exposed to chickens [23]. Of the 150 affected birds, 22 died from 1 day to 2 weeks after the arrival of clinical signs that included ruffled feathers, dyspnea, and mucus accumulation at the ricti. These birds had concomitant trichomoniasis. Eight-week-old pigeons that were infected experimentally with the agent that was isolated from the chicken flock did not develop clinical signs, whereas 4-week-old experimentally infected specific pathogen free (SPF) chickens did [23]. No further reports have been published.

Passerine birds

Antibodies to infectious bronchitis virus were demonstrated in free-ranging passerine birds in Europe [4].

Severe acute respiratory syndrome

Human severe acute respiratory syndrome (SARS) is due to a coronavirus. The virus that was isolated from palm civets differs from the human SARS by an additional 29-nucleotide sequence [24]. Recent studies showed that domestic poultry (chickens, turkeys, geese, ducks, quail) are unlikely to be a reservoir for SARS and that chickens do not play a role as amplifying hosts for SARS coronavirus [25,26].

Flaviviridae

The most notable flavivirus is West Nile virus. West Nile virus is endemic along parts of the Nile River, but in the late 1990s it was found within the eastern United States and has since spread across the United States.

Crows, jays, raptors, and horses are susceptible species, whereas poultry are considered to be resistant. There are sporadic reports of otherwise healthy psittacine birds dying of West Nile virus [27]. If people or dogs are affected, they usually are older or immunosuppressed.

The West Nile virus is spread by mosquitoes; therefore, control of mosquitoes helps to control this disease greatly. There is speculation that the disease may be transmitted horizontally; this could be true as a result of the high viral load in vascular feather pulp and potential trauma and exposure to that pulp [28].

Clinical signs range from none in resistant species (eg, poultry) to neurologic signs (ataxia, circling, head tilt, seizures) and death in susceptible species. A complete blood cell count usually is normal or a lymphocytosis is present.

A serum antibody ELISA test is available for antemortem testing. Histopathologically, lesions are observed most often in the heart, liver, kidney, and brain. In owls, lesions in the heart were most common and most severe [29]. Recently, it was shown that a reverse transcriptase PCR of vascular feather pulp was significantly more successful at detecting West Nile virus than from liver, spleen, or cloacal samples [28]. Immunohisto-chemistry also is available for evaluating formalin-fixed tissue.

Treatment consists of supportive care. Recently, in humans, the use of α-interferon seemed to result in better success. This disease already has spread throughout the United States so it is too late to prevent the disease in this country. A conditionally-licensed vaccine is available for use in horses (Fort Dodge Animal Health, Fort Dodge, Iowa); it is being used intramuscularly in birds at the same or a reduced dose [30,31]. Studies are underway to evaluate the antibody response and protectiveness of this vaccine in some species of birds. Also, a DNA plasmid vaccine that is specifically for use in raptors is being developed [32].

Herpesviridae

Herpesviruses are enveloped DNA viruses that usually are host-specific. They cause mild disease in adapted hosts, severe disease in nonadapted hosts, and life-long latent infections with intermittent shedding. Many avian herpesviruses have been described, including Pacheco's disease virus, Amazon tracheitis virus, pigeon herpesvirus, and others.

Pacheco's disease virus

At least three distinct serotypes of Pacheco's disease virus (PDV) occur worldwide. Some psittacine strains can cause clinical signs in pigeons. All psittacine birds can show clinical signs of disease from PDV. Experimental transmission of PDV was successful through the oral, intramuscular, intranasal, and intraocular routes.

Most often, birds are found dead with no premonitory signs. Occasionally, birds may exhibit depression, anorexia, diarrhea, biliverdinuria, or neurologic signs [4]. Mortality can be 18% to 80%. Outbreaks can occur in aviaries with or without recent acquisition of birds. The virus occurs subclinically in many aviaries.

The visualization of hepatic intranuclear inclusion bodies (Cowdry type A bodies) is highly suggestive of PDV. An FA test is available to diagnose PDV from liver samples. Histologically, severe hepatic necrosis is observed. Other diagnostic modalities that are available include EM, cell culture, and viral-specific DNA in situ hybridization. Antibody tests are available, but are of little practical value because antibodies to PDV develop inconsistently; therefore, a bird with no titer could be latently infected and is shedding the virus |4|. Any bird that has an antibody titer to PDV is considered to be latently infected.

The use of acyclovir, 80 mg/kg, by mouth every 8 hours for 10 days, reduces mortality during an outbreak. Herpesvirus vaccines in general, including the PDV vaccine for use in psittacine birds (Psittimune PDV, Biomune, Lenexa, Kansas), prevent severe disease but do not protect the bird from latent infections or shedding [4]. The pigeon and domestic fowl herpesvirus vaccines do not work in psittacine birds because the viruses are serologically different.

Amazon tracheitis virus

Amazon tracheitis virus (ATV) is similar to infectious laryngotracheitis in gallinaceous birds and may be a variant. The ATV was isolated from the trachea of Amazon parrots in Europe. There is a peracute, an acute, and a chronic presentation. The peracute form is associated with severe diphtheritic lesions in the trachea and bronchi. The acute form is associated with tracheal necrosis. The chronic form is associated with conjunctivitis, sinusitis, coughing, rales, and secondary bacterial respiratory infections.

Histologically, hemorrhage and necrosis of the trachea can be observed, as well as a pseudomembranous tracheitis, pharyngitis, ingluvitis, and air sacculitis. Virus isolation and EM can be performed on tracheal exudates. Supportive care and antimicrobials for secondary infections are recommended. It is unknown if infectious laryngotrachitis (ILT) vaccines that were developed for poultry would prevent ATV; however, these modified live vaccines are not recommended in psittacine birds because of the possibility of inducing disease with vaccination [4].

Parakeet herpesvirus

Parakeet herpesvirus has been described in various species in the genus Neophema in the United States and Japan. It is believed to be distinct from ILT because intranuclear inclusion bodies are observed in lungs, rather than trachea. The virus has been associated with severe respiratory disease that involves the lungs and air sacs, central neurologic signs, and death [4]. A mortality rate of 21% (14/67) was reported in *Psitticula* species that were being held at a Japanese quarantine station within 2 weeks of being imported from India [33].

Herpesvirus associated with wartlike skin lesions

A herpesvirus has been demonstrated by EM in wartlike tissue that is observed most often on the toes of cockatoos. Macaws also are affected, but their toe lesions tend to be flat, plaquelike, depigmented areas. These lesions seem to persist for years with no clinical problems; often, long-time cagemates are unaffected which suggests a low infection rate [4].

Herpesvirus of European budgerigars

This herpesvirus has been described only in budgerigars in Europe. It is considered to be serologically different from other herpesviruses of psittacine birds, but serologically related to pigeon herpesvirus. This virus may cause feather abnormalities, but mainly is known for causing early embryonic death (ie, dead in shell). It is believed to be egg transmitted [4].

Herpesvirus associated with papillomatosis (wartlike gastrointestinal lesions)

Papillomatous lesions in the gastrointestinal (GI) tract should not be confused with warts that are caused by papillomavirus (see Papillomavirnae, under Papovaviride section). Grossly, the papillomatous cloacal masses look wartlike, but histologically, they lack the long rete pegs that are associated with true papilloma lesions [4]. Recently, internal and external papillomatosis has been associated with psittacid herpesvirus–1 (PHV-1) [34,35].

Species that are affected most commonly include the Amazon parrots and macaws. Amazon parrots are prone to developing concomitant bile duct or GI tract carcinoma that recently was associated with PHV-1, genotype 3 [35]. Also an African gray parrot, an eclectus, and a cockatiel were diagnosed with PHV-1–associated liver or GI carcinoma [35]. Papillomatosis has not been reported in free-ranging birds.

Clinical signs of papillomatosis include wartlike masses anywhere along the GI tract—most commonly in the cloaca and oropharynx. Birds may exhibit weight loss, signs of straining to defecate, soiled vent, or blood in stool. Some masses cause GI obstruction with associated clinical signs. Because the virus is latent, birds that have been treated previously may have recurrence of lesions and signs with stress. Amazon parrots that have bile duct carcinoma may exhibit biliverdinuria and lethargy and bile acid levels may be high [36].

Diagnosis is suggestive based on gross appearance, location, and the tissue turning white after application of (5% acetic acid) vinegar. Definitive diagnosis is based on histology. Treatment involves removing the wartlike growth. In the author's experience, it is best to apply silver nitrate to the lesion—or half of the lesion if it circumferentially involves the cloaca—every week under anesthesia until gone. Butorphanol also is administered at 2 mg/kg, intramuscularly, one time before the procedure. The lesions can recur with stress.

Orthomyxoviridae

Influenza virus is in the Orthomyxoviridae family and consists of 3 types—A, B, and C; only type A is found in birds. There are two influenza A subtypes based on surface proteins, "H" for hemagglutinin and "N" for neuraminidase. There are 15 known H subtypes and 9 known N subtypes. All subtypes are found in birds, but only H1, H2, H3, N1, and N2 are found commonly in humans. All avian virulent strains to date have been H5 or H7 subtype, but most H5 or H7 isolates have been of low virulence. For example, an H5 or H7 automatically is called a "highly pathogenic" strain, but live chicken pathogenicity testing can later find the virus to be nonpathogenic in chickens (eg, 2003 Texas H5N2 outbreak). Further information on avian influenza can be obtained at http://www.oie.org and http://www.cdc.gov.

The disease was described first in the 1890s and was isolated first in 1955. All species of birds are considered to be susceptible to influenza A viruses, but most infections result in no to mild clinical signs. In 1997, influenza A was shown to pass from chickens to humans in an outbreak in Hong Kong in which 18 people became sick and 6 died who had contact with infected birds; this outbreak was controlled by killing 1.5 million chickens. An H5N1 subtype that was isolated from humans in Thailand and Vietnam was sequenced genetically and all genes were of bird origin. In 2003, H5N1 subtype was confirmed in turkey poults in Cambodia, China (Hong Kong had a single positive peregrine falcon), Indonesia, Japan, Laos, South Korea, Thailand and Vietnam. Also in 2003, influenza A H7N7 was isolated from poultry workers in The Netherlands. More than 80 human cases were reported and one person, a veterinarian, died.

Wild birds are the natural hosts for influenza A, especially waterfowl, and generally do not exhibit illness from infection. Domestic birds, particularly chickens, are susceptible to high mortality with influenza A. Animal to human transmission has occurred, but is unusual. The chance that influenza A will develop human genes and transmit from human to human is real and may happen in the near future. Because the virus is enveloped it is susceptible to common disinfectants.

The many different strains of influenza A create a wide variety of clinical signs. Poultry tend to have respiratory disease and high mortality. Influenza A has been recovered from psittacine birds that were asymptomatic and those that showed signs of severe respiratory disease [4].

Many tests are available, including the AGID test, ELISA, hemagglutination inhibition, and reverse transcriptase (RT)-PCR. Not all birds develop demonstrable antibodies on the ELISA.

Supportive care is recommended for treatment in humans. Birds with H5 or H7 are required to be depopulated. The 2003 influenza A H5N1 that was isolated from people in Hong Kong was sequenced genetically and was found to be resistant to two antiviral drugs, amantidine and rimantidine; however, it was sensitive to oseltamavir and zanamavir. Currently, prevention consists of testing and culling after involving the federal government.

Papovaviridae

The Papovaviridae family of nonenveloped viruses consists of two subfamilies—Papillomavirinae and Polyomavirinae. Papillomavirinae causes true warts in birds, whereas Polyomavirinae causes polyomavirus in birds.

Papillomavirinae

Papillomaviruses are highly host specific and cause benign skin masses on the face and head of African gray parrots and the legs and feet of passerine birds in the Fringillidae (ie, finches) family. Papillomatous masses that are found in the cloacal region or GI tract are not caused by a papillomavirus (see herpesvirus) [37]. Masses that are caused by papillomavirus are characterized histologically by long, thin folds of hyperkeratotic epidermis that is accompanied by acanthotic parakeratosis (rete pegs) [4]. The papillomavirus that was isolated from African gray parrots was related antigenically to bovine papillomavirus type 1 [38]. Differential diagnosis is based on demonstration of intranuclear inclusion bodies with 45-nm to 50-nm viral particles by way of EM.

Polyomavirinae

Polyomaviruses are host specific and cause subclinical disease in mammals; however, in psittacine and other birds they cause severe clinical disease. The first polyomavirus that was described in any animal was in budgerigars in 1981 [39,40].

The disease in budgerigars is different than what is described for other psittacine birds and is called budgerigar fledgling disease. Budgerigars exhibit poorly developing feathers, especially the contour feathers; abdominal distension; subcutaneous hemorrhage; and the young can die acutely with or without neurologic signs, such as ataxia and tremors.

Psittacine birds, other than budgerigars, usually develop subclinical disease because their mature, healthy immune system prevents the acute form of the disease. These birds shed polyomavirus intermittently for the remainder of the their life. Immature nonbudgerigar psittacine birds can develop peracute, acute, or chronic forms of the disease that usually depends on age at exposure. Peracute death occurs in young birds. Acute disease also occurs in young birds with an approximate mortality rate of 27% to 41% and includes 12 to 48 hours of depression, anorexia, delayed crop emptying, regurgitation, diarrhea, dehydration, subcutaneous hemorrhage, dyspnea, and polyuria. The subcutaneous hemorrhages are seen most easily over the crop, carpi, or cranium. One outbreak that was described is typical of an acute event; 31% of birds between 28 and 48 days of age died within a 6-week period.

Chronic disease is associated with weight loss, intermittent anorexia, polyuria, renal failure, poor feather growth and signs of immunosuppression, such as secondary bacterial or fungal infections. It seems that the chronic disease occurs in immunosuppressed adults that cannot overcome the clinical signs of infection.

Transmission of polyomavirus is through exposure to excretion and secretions, especially urine. Vertical transmission has been proven in budgerigars and is suspected in other species. The incubation period is believed to be between 2 and 14 days.

Gross necropsy findings include ascites, hepatosplenomegaly, renomegaly, pale cardiac or skeletal muscle tissue, feather dystrophy, and subcutaneous hemorrhages. Histologic findings include hepatic necrosis, karyomegaly in hepatic and splenic tissue, bursal lymphoid depletion, and membranous glomerulopathy [4]. Viral intranuclear inclusion bodies can be detected in feather follicles and renal tissue. A DNA probe (PCR) test is available to detect viral DNA in tissue or feces. Antibody tests are available; a positive result denotes that exposure has occurred and that the bird probably sheds virus intermittently. Many birds in aviaries are affected subclinically and are a constant source of infection for other birds in the aviary and are a particular danger for young birds; therefore, all birds should be vaccinated.

Psittacine birds should be vaccinated with the commercially available, fully licensed vaccine Psittimune (Biomune). This inactivated vaccine is administered subcutaneously and has been proven to be safe and effective [41,42]. Generally, young birds are vaccinated starting at 5 weeks of age and are protected optimally at 2 weeks after the second vaccine. Birds that weigh 200 g or more receive 0.5 mL, whereas birds that weigh less than 200 g receive 0.25 mL of vaccine.

There is no treatment for the disease. The prognosis is grave if clinical signs are present. Because it is a nonenveloped virus it is stable in the environment and difficult to destroy. A synthetic phenol, sodium hypochlorite (bleach), stabilized chlorine dioxide, and 70% ethanol have shown some efficacy in destroying the virus. A swab of the environment can be tested with a DNA probe to evaluate the disinfection process.

Polyomavirus in passerine birds

This disease was described first in Canada in 1986; 51% of 2- to 3-day-old chicks died acutely after 0 to 48 hours of nonspecific illness. Acute mortality can be observed in passerine birds of any age. Those that survive can develop feather dystrophy and deformed beaks. Gross necropsy findings include hepatomegaly with paleness and mottling, splenomegaly with congestion, and subserosal intestinal hemorrhage [4]. Intranuclear inclusion bodies can be observed in the spleen, bone marrow, intestine, kidney, heart, and liver. DNA probe tests that detect polyomavirus in psittacine birds do not detect polyomavirus in passerine birds; therefore, one must request a specific passerine bird polyomavirus DNA probe test from the laboratory [43,44].

Paramyxoviridae

Newcastle's disease virus

Newcastle's disease is caused by a paramyxovirus of which there are four pathotypes—lentogenic, mesogenic, neurotropic velogenic, and viscerotropic velogenic. The last also is known as exotic Newcastle's disease, which is a foreign animal disease in the United States. Exotic Newcastle's disease is highly pathogenic in poultry, but psittacine birds and pigeons are more resistant to infection. Free-ranging pigeons can serve as mechanical vectors. The nonexotic forms of Newcastle's disease used to be common before the use of vaccines.

The disease is spread by direct contact with viral particles from aerosolization of respiratory secretions or feces or from food, water, or litter that is contaminated with feces. Recovered birds are believed to shed the virus indefinitely. The incubation period ranges from 3 to 28 days, but the average is 5 days in chickens.

The clinical signs vary with the pathotype but include depression, diarrhea, anorexia, ruffled feathers, oculonasal discharge, conjunctivitis, dyspnea, ataxia, muscle tremors, paralysis, and death. In people, Newcastle's disease causes a mild, acute granular conjunctivitis, general malaise, and sinusitis that resolve within 7 to 20 days.

Avian serum can be sent to the San Bernardino County Laboratory in California for antibody testing for paramyxovirus (PMV)-1, -2, and -3. Many tests are available, including the HI test, ELISA, AGID, and virus isolation. Quarantine stations were developed to keep exotic Newcastle's disease out of the United States; therefore, any bird that enters U.S. quarantine is tested serologically.

Pet birds that have a nonexotic form of Newcastle's disease should be offered supportive care. Any bird, pet, that is positive for exotic Newcastle's disease in the United States is to be reported to federal authorities and destroyed.

Vaccines are available for control of nonexotic forms of Newcastle's disease in the United States. Exotic Newcastle's disease is controlled by the federal government by test and cull methods. A vaccine is available in Europe for exotic Newcastle's disease, but its use is not allowed in the United States because the tactic to control the disease in this country is to test and cull any positive birds.

Paramyxovirus (PMV)-2

Passerine birds that are infected with PMV-2 show mild, self-limiting disease, whereas psittacine birds, especially African gray parrots, that are infected with the same virus have severe difficulty breathing, diarrhea, and high mortality [4]

Paramyxovirus (PMV)-3

PMV-3 has been isolated from a large variety of psittacine and passerine birds and from clinically normal and dead birds. In *Neophema* spp the morbidity is high and the mortality is low. All ages are affected, but neonates have the most severe symptoms. Clinical signs include torticolis and ataxia. Gross necropsy findings include pulmonary edema and congestion, hepatomegaly, and pancreatic atrophy [4].

Nestling cockatiels had high mortality after clinical signs of opisthotonus, leg paralysis, and dyspnea. Gross necropsy findings included cardiomegaly and pericardial effusion. Virus was recovered form the brain and heart. Moluccan cockatoos that had PMV-3 presented with ataxia and death. Intracytoplasmic inclusion bodies were seen in the brain. Finches that had PMV-3 exhibited conjunctivitis that was followed by anorexia, diarrhea, and dyspnea. Only two of five finches that were infected experimentally died in one study [4].

Paramyxovirus (PMV)-5

PMV-5 has been isolated from budgerigars in Japan that exhibited acute clinical signs including depression, diarrhea, torticolis, dyspnea, and high (90%) mortality. Young birds seem to be more susceptible than adults [4].

Paramyxovirus (PMV) of columbiformes

This pigeon strain of PMV causes severe neurologic disease in pigeons, but only mild disease in chickens. Young pigeons suffer a high mortality rate of up to 90%. An inactivated subcutaneous vaccine against pigeon PMV-1 is used extensively in Europe. It is recommended to vaccinate pigeons at 3 to 4 weeks of age and again at 4 months before breeding; also, all new pigeons should be vaccinated [4].

Poxviridae

Poxviruses are the largest of viruses and the genus Avipoxviruses are found worldwide in greater than 20 families of birds. There are many species of Avipoxvirus, such as psittacine pox, canary pox, pigeon pox, falcon pox and fowl pox. Each species of pox has varied host specificity, but typically the most severe clinical signs are seen in its natural host.

The virus is transmitted by way of mosquito or by way of mechanical means through broken skin. Mites and other blood-feeding insects have been implicated in the spread of disease. The incidence of disease increases with rainfall and mosquito population. Mosquitoes were shown to remain infective for weeks to months. The virus is extremely durable in the environment; it is killed by 1% KOH, 2% NaOH, or 5% phenol, or heating to 50°C for 30 minutes or 60°C for 8 minutes [4].

Ten to 14 days postinfection, birds can show blepharitis, ocular discharge, rhinitis, and conjunctivitis that is associated with raised papules. Later, ulcerations of the lid margins occur that dry to crusty lesions, scabs, and eventually, scars. Persistent infections of 13 months or longer have been reported in chickens. Stress can activate a latent infection.

Clinical signs can be divided into "dry" pox that consists of cutaneous papular lesions, and "wet" pox that consists of mucosal papular lesions of the oropharynx. Occasionally, birds may display neurologic signs.

Diagnosis is based on typical clinical signs and histologic finding of Bollinger bodies (intracytoplasmic inclusion bodies) in skin or mucosal cells, and is considered to be pathognomonic.

Treatment consists of providing supportive care. Scabs should be left to heal naturally to lessen scarring. Vaccines have been created for chickens, pigeons, turkeys, canaries, quail, waterfowl, falcons, and Amazon parrots. The fowl pox vaccine may provide some protection in pigeons, waterfowl, and falcons. It is recommended to vaccinate before the breeding or mosquito season. Maximum protection occurs 3 to 4 weeks after vaccination.

Psittacine pox

This disease used to be common in recently imported Amazon parrots—especially blue-fronted Amazon parrots, macaws, and pionus—but is seen rarely today. Occasionally, an older imported bird will present with old pox scars on the eyelids, nostrils, and face. Corneal ulcers or crystallization may occur with or without a uveitis, that also can lead to later scarring [45]. In one outbreak of yellow-headed Amazon parrots, 28% died that developed the "wet" or mucosal form of pox that can cause pneumonia or air sacculitis [46].

Agapornis pox

Lovebirds have their own species of pox—Agapornis pox—which causes morbidity and mortality rates of approximately 75%. Lovebirds develop areas of dry, darkened skin, but not papules.

Canary pox

In canaries, aggression brings about wounds near the carpi; therefore, pox nodules often are seen there. Mortality rates can reach 70% or more as the result of a septicemic form that occurs commonly in canaries [4]. Also, lung tumors can form later in canaries. It is recommended to vaccinate with Poximune C (Biomune) at fledging stage (>4 weeks) and 1 month before breeding season. This is a wing web vaccine. Some canary breeders vaccinate every 6 months. Vaccinating healthy canaries during an outbreak is acceptable if care is taken not to become a mechanical vector while vaccinating.

Pigeon pox

Mortality in squab may reach 50%. Cutaneous tumors can form later in Columbiforme birds. The available vaccine is attenuated and is applied to racing pigeons by plucking feathers on the leg and brushing the vaccine on the open follicles. This should produce a swelling and a yellowish-brown discoloration in the area in 5 to 7 days [4].

Poxvirus of wild passerine birds

Many veterinarians who practice avian medicine take in wildlife. It is important to realize that passerine pox may enter your practice by way of the commonly infected and commonly presented wild house finch.

Reoviridae

The Reoviridae family consists of three genera—Orthoreovirus, Orbivirus, and Rotavirus—all of which have been described in birds. Orthoreoviruses are described more commonly in birds than the other genera of viruses; at least 11 antigenically distinct serotypes have been described in psittacine birds, pigeons, domestic fowl, geese, and raptors. Orthoreoviruses have been studied best in chickens and cause tenosynovitis. In other birds, the clinical signs are varied and can be vague and complicated by concomitant bacterial or fungal infections. Transmission mainly is by way of the fecal oral route, but also occurs from exposure to respiratory sections. The virus is stable in the environment and is difficult to destroy.

In psittacine birds, the liver is the primary organ that is targeted; clinical signs include anorexia, depression, weight loss, subcutaneous hemorrhage, diarrhea, ascites, dyspnea, nasal discharge, ataxia, paralysis, uveitis, hypopyon, and edema of the head and legs [4]. African gray parrots and cockatoos that were infected experimentally died or showed no clinical signs [47]. In another study, antibodies to reovirus were detected in 4.1% of psittacine birds in a U.S. quarantine facility over a 7-year period [4]. Recently, reovirus was described in African gray parrots and budgerigars; many had concomitant Pacheco's disease virus or fungal infections and lymphoid depletion [48,49]. Reovirus probably is more prevalent than currently is believed; it is likely that more information will be forthcoming as it is studied further.

Diagnosis is based on EM and is confirmed by virus isolation. Because the virus can be present with no clinical signs and concomitant disease is common, the presence that is indicated by EM is not enough to reach a definitive diagnosis. Intracytoplasmic inclusion bodies may be observed and many other tests have been used in poultry, including AGID, virus neutralization, ELISA, and immunofluorescent antibody (IFA). Virus has been detected in bursa, spleen, pancreas, and even in circulating cells [49]. The vaccine that is available for use in chickens is not recommended for use in psittacine birds because the chicken and psittacine reoviruses are not related antigenically [4].

Pigeon reovirus

Reportedly, 8% to 16% of homing pigeons in Europe had reovirus antibodies [4]. Clinical signs in pigeons include diarrhea and dyspnea and signs that are consistent with hepatitis.

Retroviridae

Retroviruses cause lymphoid tumors, including erythroid leukosis and myleoblastosis, in chickens. Clinical lesions that are similar to those described in chickens have been described in many other species of birds, including canaries, psittacine birds, and pigeons; however, a definitive causal relationship between retroviruses and lesions in these birds has not been proven [4]. Canaries that have lymphoid leukosis often have masses about the head and neck.

Retroviruses can be transmitted vertically and horizontally. Diagnostic modalities include complement fixation (CF) and ELISA to detect virus, and VN and ELISA to detect antibody. The virus is unstable outside of the host and is destroyed easily by common disinfectants. There is no vaccine available; therefore, hygiene is of the utmost importance.

Rhabdoviridae

All warm-blooded animals, including birds, are susceptible to rabies virus. Avian infections are rare, but have been reported in chickens, waterfowl, and raptors. A passive hemagglutination test in wild birds was seropositive in 23.1% of 65 predatory birds and 2.9% of 278 nonpredatory birds. Although no case of avian-to-human transmission of the rabies virus has been documented, pre-exposure rabies prophylaxis and preventing trauma while handling raptors is recommended [4].

Togaviridae

The Togaviridae family consists of three genera. Only the alphaviruses are described in birds, including Eastern equine encephalomyelitis (EEE) and Western equine encephalomyelitis (WEE).

Birds are a known reservoir for EEE and WEE; many species of birds are involved. The virus is transmitted through a mosquito bite. In pheasants, it was shown to be transmitted through the broken skin that is created by pecking.

The mosquitoes that are known to transmit EEE virus in the United States and Canada include *Culiseta melanura*, *Aedes* spp, and *Coquillettidia* spp. *Culiseta melanura* feeds mainly on birds and rarely on people, whereas *A sollicitans* and *A vexans* feed on birds and people, but are less likely to be infected with the virus. Bird mites are known to be potential vectors.

The mosquitoes that are known to transmit WEE virus include *Culex tarsalis* in the western United States and Canada and *Culiseta melanura* in the eastern United States. People are an accidental host to the virus. If infected with EEE, the mortality rate is up to 80%; if infected with WEE, the mortality rate ranges from 5% to 15%. In Canada, a direct connection was shown between the prevalence of the virus in wild birds and the number of human cases. During the 1962 outbreak in Saskatchewan, Canada, the WEE virus was isolated from 22% of the wild bird population.

Birds may show no clinical signs or ataxia, trembling, weakness, paralysis, and death. People show varying degrees of neurologic signs and death. Antemortem diagnosis is difficult. Histopathologic lesions of lympho-plasmacytic encephalitis is suggestive. There is no specific treatment for EEE and WEE other than supportive care [4].

Proventricular dilatation disease

The causative organism of this disease has been identified as an 89-nm virus that is unclassified. The route of transmission is by way of fecal to oral and seems to affect birds of many orders, including psittacine birds.

Clinical signs include severe, chronic weight loss; regurgitation; delayed crop emptying; ravenous appetite; undigested food in stool; and neurologic signs (eg, falling off perch). The virus paralyzes the nerves in the proven-triculus and the bird essentially starves to death—despite a good appetite—because of its inability to process food. Other nerves or organs can be affected alone or in conjunction with the proventriculus.

Suggestive diagnostic testing includes radiographs that demonstrate proventricular dilitation and whole, undigested food particles or seeds in the feces. Many diseases can cause proventricular dilation, including disease from parasites, yeast, megabacterium, mycobacterium, foreign body, neo-plasia, and lead and zinc toxicosis.

Definitive diagnostic testing includes crop biopsy that demonstrates a "lymphoplasmocytic ganglioneuritis." EM that demonstrates a 89-nm virus in the feces is highly suggestive, but the virus is labile and does not withstand overnight shipping.

Birds usually die within 2 years of developing clinical signs. Recently, treatment with cyclo-oxygenase (COX)-2 inhibitors significantly improved clinical signs; however, the mechanism against the virus is unknown [50]. Prevention consists of avoiding exposure to known infected birds.

Summary

Many viruses definitively cause disease in our companion birds, whereas other viruses have been implicated or associated with typical clinical signs. Some families of viruses that have been discovered in mammals have not been associated with disease in birds. It is imperative to perform a necropsy on any birds that die—whether a pet, aviary, or display bird, and despite the fact that other diseases may be present—because viruses can occur concurrently, especially when immunosuppression is present. Also, it is imperative to use available vaccines to decrease and control the incidence of these diseases, as has occurred in the canine and feline pet populations.

References

[1] Balamurugan V, Kataria JM. The hydropericardium syndrome in poultry-a current scenario. Vet Res Commun 2004;28(2):127–48.

[2] International Committee on Taxonomy of Viruses. Available at: http://www.virology.net. Accessed August 29, 2004.

[3] Soike D, Hess M, Prusas C, et al. Adenovirus infektionen in papageien [Adenovirus infections in psittacines] Tierarztl Prax Ausg K Klientiere Heimtiere 1998;26(5):354–9 [in German].

[4] Ritchie BW. Avian viruses, function and control. Lake Worth (FL): Wingers Publishing; 1995.

[5] Goodwin MA, Davis JF. Adenovirus particles and inclusion body hepatitis in pigeons. J Assoc Avian Vet 1992;6:37–9.

[6] Goryo M, Ueda Y, Umemura T, et al. Inclusion body hepatitis due to adenovirus in pigeons. Avian Pathol 1988;17:391–401.

[7] Ramis A, Marlasca MJ, Majo N, et al. Inclusion body hepatitis (IBH) in a group of eclectus parrots (*Eclectus roratus*). Avian Pathol 1992;21:165–9.

[8] Ramis A, Latimer KS, Niagro FD, et al. Diagnosis of psittacine beak and feather disease (PBFD) viral infection, avian polyomavirus infection, adenovirus infection and herpesvirus infection in psittacine tissues using DNA in situ hybridization. Avian Pathol 1994;23: 643–57.

[9] Latimer KS, Niagro FD, Williams OC, et al. Diagnosis of avian adenovirus infection using DNA in situ hybridization. Avian Dis 1997;41(4):773–82.

[10] Raue R, Hafez HM, Hess M. A fiber gene based polymerase chain reaction for specific detection of pigeon adenovirus. Avian Pathol 2002;31(1):95–9.

[11] Ritchie BW, Niagro FD, Lukert PD, et al. Characterization of a new virus from cockatoos with psittacine beak and feather disease. Virology 1989;171:83–8.

[12] Woods LW, Latimer KS, Niagro FD, et al. A retrospective study of circovirus infection in pigeons: nine cases (1986–1993). J Vet Diagn Invest 1994;6:156–64.

[13] Greenacre CB, Latimer KS, Niagro FD, et al. Psittacine beak and feather disease in a scarlet macaw (*Ara macao*). J Assoc Avian Vet 1992;6:95–8.

[14] Schoemaker NJ, Dorrestein GM, Latimer KS, et al. Severe leukopenia and liver necrosis in young African grey parrots (*Psitticus erithacus erithacus*) infected with psittacine circovirus. Avian Dis 2000;44(2):470–8.

[15] Ritchie BW. Management of common avian infectious diseases. Proceedings of the Annual Western Veterinary Conference. Las Vegas (NV): Western Veterinary Conference; 2003. p. 1–9.

[16] Soike D, Albrecht K, Hatterman K, et al. Novel circovirus in mullard ducks with developmental and feathering disorders. Vet Rec 2004;154(25):792–3.

[17] Heath L, Martin DP, Warburton L, et al. Evidence of unique genotypes of beak and feather disease virus in southern Africa. J Virol 1994;78(17):9277–84.

[18] Shivaprasad HL, Chin RP, Jeffrey JS, et al. Particles resembling circovirus in the bursa of fabricius of pigeons. Avian Dis 1994;38:635–41.

[19] Smyth JA, Weston J, Moffett DA, et al. Detection of circovirus infection in pigeons by in situ hybridization using cloned DNA probes. J Vet Diagn Invest 2001;13(6):475–82.

[20] Roy P, Dhillon AS, Lauerman L, et al. Detection of pigeon circovirus by polymerase chain reaction. Avian Dis 2003;47(1):218–22.

[21] Hirai K, Hitchner SB, Calnek BW. Characterization of a new coronavirus-like agent isolated from parrots. Avian Dis 1978;23:515–25.

[22] Hirai K, Hitchner SB, Calnek BW. Correction in identification of a new coronavirus-like agent isolated from parrots. Avian Dis 1982;26(1):169–70.

[23] Barr DA, Reece RL, O'Rourke D, et al. Isolation of infectious bronchitis virus from a flock of racing pigeons. Aust Vet J 1988;65:228.

[24] Webster RG. Wet markets—a continuing source of severe acute respiratory syndrome and influenza? Lancet 2004;363(9404):234–6.

[25] Swayne DE, Suarez DL, Spackman E, et al. Domestic poultry and SARS coronavirus, southern China. Emerg Infect Dis 2004;10(5):914–6.

[26] Weingartl HM, Copps J, Drebot MA, et al. Susceptability of pigs and chickens to SARS coronavirus. Emerg Infect Dis 2004;10(2):179–84.

[27] Tully TN, Nevares J, Diaz O, et al. Determining the seroprevalence of West Nile Virus in an exposed psittacine population. Proceedings of the Annual Conference of the Association of Avian Veterinarians. Pittsburgh (PA): Association of Avian Veterinarians; 2003. p. 17–8.

[28] Docherty DE, Long RR, Griffin KM, et al. Corvidae feather pulp and West Nile Virus detection. Emer Infect Dis 2004;10(5):907–9.

[29] Fitzgerald SD, Patterson JS, Kiupel M, et al. Clinical and pathologic features of West Nile Virus infection in native North American owls (Family Strigidae). Avian Dis 2003;47:602–10.

[30] Olsen GH, Miller KJ, Docherty D, et al. West Nile Virus vaccination and challenge in sandhill cranes (*Grus canadensis*). Proceedings of the Annual Conference of the Association of Avian Veterinarians. Pittsburgh (PA): Association of Avian Veterinarians; 2003. p. 123–4.

[31] Llizo SY. Management of West Nile Virus in zoo birds. Proceedings of the Annual Conference of the Association of Avian Veterinarians. Pittsburgh (PA): Association of Avian Veterinarians; 2003. p. 117–122.

[32] Redig P, Tully T, Ritchie BW, et al. Testing of a DNA-plasmid vaccine in red-tailed hawks (*Buteo jamaicensis*). Proceedings of the Annual Conference of the Association of Avian Veterinarians. New Orleans (LA): Association of Avian Veterinarians; 2004. p. 29–31.

[33] Tsai SS, Park JH, Hirai K, et al. Herpesvirus infections in psittacine birds in Japan. Avian Pathol 1993;22:141–56.

[34] Goodwin MA, McGee ED. Herpes-like virus associated with a cloacal papilloma in an orange-fronted conure (*Aratinga canicularis*). J Assoc Avian Vets 1993;7:23–6.

[35] Styles DK, Tomaszewski EK, Phalen DN. Psiattacid herpesvirus associated with internal papillomatous disease in psittacine birds. Proceedings of the Annual Conference of the Association of Avian Veterinarians. New Orleans (LA): Association of Avian Veterinarians; 2004. p. 79–81.

[36] Hillyer EV, Moroff S, Hoefer H, et al. Bile duct carcinoma in two out of ten Amazon parrots with cloacal papillomas. J Assoc Avian Vets 1991;5:91–5.

[37] Latimer KS, Niagro FD, Rakick PM, et al. Investigation of parrot papillomavirus in cloacal and oral papillomas of psittacine birds. Vet Clin Pathol 1997;26(4):158–63.

[38] Jacobson ER, Mladinich CR, Clubb S. Papilloma-like virus infection in an African grey parrot. J Am Vet Med Assoc 1983;183:1307–8.

[39] Davis RB, Bozeman LH, Gaudry OJ, et al. A viral disease of fledgling budgerigars. Avian Dis 1981;25:179–83.

[40] Bozeman LH, Davis RB, Gandry D, et al. Characterization of a papovavirus isolated from fledgling budgerigars. Avian Dis 1981;25:972–80.

[41] Ritchie BW, Vaughn SB, Leger JS, et al. Use of an inactivated virus vaccine to control polyomavirus outbreaks in nine flocks of psittacine birds. J Am Vet Med Assoc 1998;21(5):685–90.

[42] Ritchie BW, Latimer KS, Leonard J, et al. Safety, immunogenicity, and efficacy of an inactivated avian polyomavirus vaccine. Am J Vet Res 1998;59(2):143–8.

[43] Garcia AP, Latimer KS, Niagro FD, et al. Avian polyomavirus infection in three black-bellied seed crackers (*Pyrenestes ostrinus*). J Assoc Avian Vet 1993;2:79–82.

[44] Garcia AP, Latimer KS, Niagro FD, ct al. Diagnosis of polyomavirus infection in seed crackers (*Prenestes* sp.) and blue bills (*Spermophaga haematina*) using DNA in situ hybridization. Avian Pathol 1994;23:525–37.

[45] Karpinski LG, Clubb SL. Post pox ocular problems in blue-fronted Amazon and blue-headed pionus parrots. Proc Annu Conf Assoc Avian Vet; 1985. p. 91–100.

[46] Boosinger TR, Winterfield RW, Feldman DS, et al. Psittacine pox virus: virus isolation and identification, transmission, and cross-challenge studies in parrots and chickens. Avian Dis 1982;26:437–44.
[47] Graham DL. Characterization of a reo-like virus and its isolation from and pathogenicity for parrots. Avian Dis 1987;31:411–9.
[48] Manvell R, Gough D, Major N, Fouchier RA. Mortality in budgerigars associated with a reovirus-like agent. Vet Rec 2004;154(17):539–40.
[49] Sanchez-Cordon PJ, Hervas J, Chacon de Lara F, et al. Reovirus infection in psittacine birds (*Psitticus erithacus*): morphologic and immunohistochemical study. Avian Dis 2002;46(2): 485–92.
[50] Dahlhausen R, Aldred S, Colaizzi E. Resolution of clinical proventricular dilatation disease by cyclooxygenase 2 inhibition. Proc Annu Conf Assoc Avian Vet. Monterey (CA); 2002. p. 9–12.

ELSEVIER
SAUNDERS

Vet Clin Exot Anim 8 (2005) 107–122

VETERINARY
CLINICS
Exotic Animal Practice

Small mammal virology

Corinna Kashuba, DVM, Charlie Hsu, VMD,
Aric Krogstad, DVM, Craig Franklin, DVM, PhD,
DACLAM*

*Department of Veterinary Pathobiology, College of Veterinary Medicine, University of
Missouri–Columbia, 1600 E. Rollins Columbia, MO 65211, USA*

This article is intended to provide information on clinically relevant viral diseases of various small mammal species encountered in veterinary practice, including mice, rats, guinea pigs, hamsters, gerbils, chinchillas, prairie dogs, hedgehogs, and sugar gliders. The number of known viral diseases varies enormously with each species, and the majority of viral infections in small mammals are asymptomatic. When researching viral infections of small mammals, one should be aware that much of the literature discusses experimental infections that may result in clinical signs and lesions that do not occur with natural infections. For details of virus biology and experimental studies involving small mammal viruses, the reader is referred to a number of excellent textbooks (see Further Readings).

Most viral infections are transient and asymptomatic. Certain predisposing factors may influence the development of clinical signs due to viral infections. The immune status of the animal is important because very young (<2 weeks old), aged, and other immunocompromised animals are at increased risk of developing clinical disease. Some diseases manifest at weaning age with the waning of maternally derived passive immunity and the relatively naïve juvenile immune system. Ongoing infections (whether bacterial or viral) predispose an animal to developing additional infections, viral or otherwise. Additionally, good husbandry is vital to small mammal health, and suboptimal husbandry can influence susceptibility to viral diseases. High ammonia levels caused by infrequent cage cleanings or overcrowding, high humidity, and poor nutrition contribute to poor health and the development of clinical viral infections. Stress factors such as

---

This work was supported in part by National Institutes of Health grant T32RR07004 and a gift from Merck, Inc.

* Corresponding author.
E-mail address: franklinc@missouri.edu (C. Franklin).

shipping and temperature fluctuations may also influence the development of clinical disease.

Because viral infections are often transient, treatment is generally confined to supportive care if the disease is not complicated by bacterial infections. If secondary bacterial infections are suspected, antibiotic therapy may also be appropriate, but careful consideration should be made when choosing antibiotics for certain antibiotic-sensitive species (eg, guinea pigs). In colony or group-housed settings, viral infections may be very contagious and difficult to eliminate if naïve animals are added or produced through breeding. In these cases, a stop breeding or "burnout" program may be effective to eliminate viral infections. In these programs, breeding is stopped and no new additional animals are added (closed colony) for 6 to 8 weeks. This allows infected animals to develop a sufficient immune response to clear infections and ultimately eliminate the virus from the colony. Quarantine of potentially exposed animals and culling of carriers may also be warranted.

Diagnosis of viral infections is often presumptive and based on clinical signs if present. Serologic testing for most rodent viral agents is available at commercial laboratories, but is often not practical for diagnosis in individual animals because antibodies develop as disease is waning or after signs have abated and the virus has been cleared. However, serologic testing may be warranted in colony settings if exposed animals need to be identified. Additionally, polymerase chain reaction (PCR)-based tests have been developed for the diagnosis of many small mammal infections.

Clinical signs in small mammals

Rodents and other small mammals are active and curious animals, especially at night. Diseased animals are inactive and exhibit hunched postures and poorly groomed, ruffled hair coats (Fig. 1). This collection of clinical signs is referred to as *sick rodent syndrome*. Diseased rats may also accumulate crusty, brick red, porphyrin secretions from the Harderian gland around the eyes, nose, and forepaws. This accumulation of "red tears" is termed *chromodacryorrhea*. Labored breathing is another sign of poor health and may indicate respiratory disease or systemic disease such as bacterial septicemia or trauma.

Viral infections of rats

Respiratory disease in rats

"Respiratory disease of indeterminate nature" is a common complaint among rat breeders and fanciers. Respiratory diseases are often multifactorial, involving pathogens such as *Mycoplasma pulmonis,* cilia-associated respiratory bacillus, rat coronaviruses, and paramyxoviruses. Viral infections

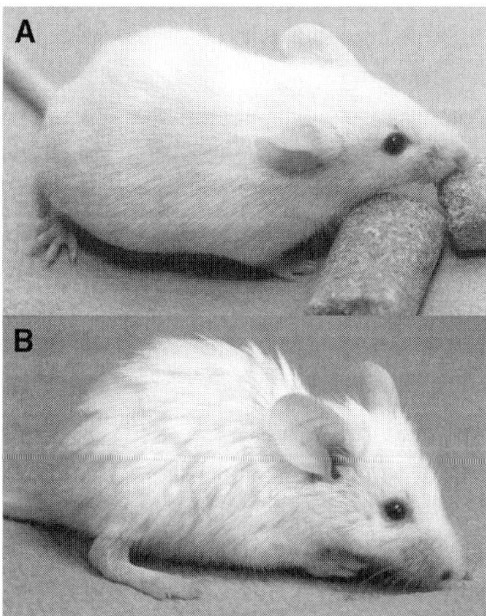

Fig. 1. Photograph of (*A*) a healthy mouse demonstrating normal groomed haircoat and (*B*) a mouse exhibiting signs of "sick mouse syndrome," including rough hair coat and hunched posture.

alone rarely result in respiratory disease, but may contribute to the establishment of more clinically relevant bacterial infections.

Rat coronaviruses, including the often-mentioned variant sialodacryoadenitis virus (SDAV), are capable of inducing salivary and lacrimal gland (especially Harderian gland) disease or respiratory disease [1]. Disease presentations range from subclinical to overt. Early clinical signs may include blepharospasm and epiphora, but often the most profound sign is intermandibular swelling due to edema and inflammation of the submandibular salivary glands. Exophthalmus and chromodacryorrhea result from swelling of the Harderian gland, and unilateral or bilateral glaucoma/ megaloglobus, hyphema, and corneal ulceration can occur secondarily. SDAV infection may potentiate clinical disease caused by *M pulmonis* or cilia-associated respiratory bacillus infections. Additionally, there are reports that SDAV infection may be associated with clinically significant reproductive disorders or decreased litter production, but this relationship remains unclear [2–4].

Pneumonia virus of mice and Sendai virus are paramyxoviruses that, as single agents, cause asymptomatic transient infections in rats. However, either virus may contribute to the severity of disease associated with bacterial respiratory pathogens, most notably *M pulmonis*.

Enteric disease in rats

Infectious diarrhea of infant rats, caused by an atypical rotavirus, was identified as the causative agent of a spontaneous outbreak of diarrhea in suckling rats in 1984 [5]. Rats orally inoculated with rotavirus isolated from the spontaneous outbreak produced signs similar to those of natural infection, including diarrhea of 5 to 6 days' duration, growth retardation, and drying and flaking of the skin. Rats more than 2 weeks of age were generally considered resistant to clinical disease [5]. This disease has been experimentally replicated numerous times; however, no additional spontaneous outbreaks have been reported, thus its significance in the pet rat population is unknown.

Systemic disease in rats

Infections with Kilham's rat virus, a parvovirus, are most often asymptomatic. However, clinical disease has been reported on two occasions. In 1966, Kilham and Margolis [6] reported fetal resorption in pregnant females, runting and hepatitis in suckling rats, and the development of ataxia and cerebellar hypoplasia in a juvenile rat assumed to be infected at birth. In 1983, Coleman and colleagues [7] reported death, dyspnea, ruffled haircoats, hunched posture, lethargy, muscular weakness, weight loss, dehydration, swollen abdomens, and cyanotic scrotums in a naïve colony of rats that became naturally infected with Kilham's rat virus. At necropsy, hemorrhage and necrosis of the brain, testes, and epididymides were observed.

Other viral infections of rats

Rats have been reported to seroconvert to a number of other viruses; however, these infections rarely, if ever, result in the development of clinical disease. These include rat parvovirus, rat minute virus, H-1 virus, rat cardiovirus, reovirus, and adenovirus.

Viral infections of mice

Dermal disease in mice

Ectromelia virus is the causative agent of mousepox. The prevalence of this virus is very low in North America [8], but when present can spread rapidly with severe consequences. The disease was termed *infectious ectromelia* because of the high incidence of limb amputation due to self-trauma in surviving mice. Clinical signs in adult mice include foot swelling, lethargy, depression, and sudden death [9], with pocks occurring most commonly on the face, muzzle, feet, or abdomen [10]. Gross pathologic changes may include small intestinal mucosal erosions, necrosis and sloughing of distal portions of the tail and limbs, or massive splenic, lymph node, thymic, and hepatic necrosis [9]. Although the prognosis for mousepox is variable, mice generally recover completely from infection and do not serve as carriers [8].

Enteric disease in mice

Mouse hepatitis virus includes a number of strains of coronaviruses that vary in pathogenicity, organ tropism, and clinical disease presentation. Infections are most often asymptomatic, but can also manifest as diarrhea, sick rodent syndrome, decreased reproductive indices, and death [11]. Gross lesions include flaccid intestines with watery intestinal contents and multifocal hepatic necrosis (Fig. 2). All ages and strains of mice are susceptible to infection, but the severity of disease is inversely correlated to age. Neonatal mice (<7 days of age) can develop marked necrotizing enterocolitis with high mortality. Adults develop minimal lesions but are equally susceptible to viral infection, propagation, and shedding [12].

Naturally occurring clinical infections due to mouse rotavirus, the causative agent of epizootic diarrhea of infant mice, have not been reported since 1948 [12–14]. Those reports described high infant mortality with yellow feces staining the perineum, dehydration, cyanosis, and fine, dry scales appearing over the shoulders and dorsum [14]. There is controversy over whether the original outbreaks involved solely rotavirus infection or represented multifactorial disease, because many clinical signs—such as high infant mortality—have not been observed in more recent mouse rotavirus infections.

Respiratory disease in mice

Infection with Sendai virus, a labile but highly contagious parainfluenza virus, is usually clinically inapparent. There have been several reports of clinical disease in mice due to Sendai virus [15–18]; however, these reports are more than 20 years old, and it is possible that multiple factors contributed to the occurrence of clinical disease in these populations. Sendai

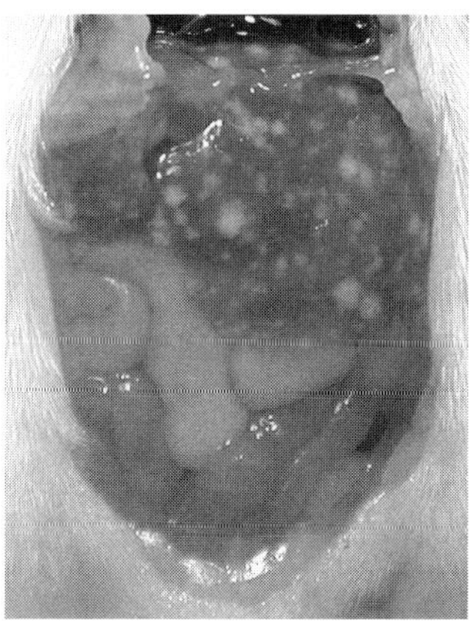

Fig. 2. Photograph of mouse infected with mouse hepatitis virus demonstrating multifocal hepatic necrosis and flaccid, fluid-content filled intestines.

virus infection will often predispose mice to clinical respiratory disease when complicated by *M pulmonis*. Antibiotic treatment may be helpful in such mixed infections.

Systemic disease in mice

Because of its polytropic nature, ectromelia virus can affect multiple systems in mice, causing conjunctivitis, blepharitis, and visceral lesions in addition to the more obvious dermal signs described previously. Polytropic strains of mouse hepatitis virus may also result in systemic disease because of their propensity to replicate in endothelium; however, this disease presentation rarely occurs in natural infections.

Neoplastic disease in mice

Murine leukemia viruses and murine mammary tumor viruses are oncogenic retroviruses of mice. Most retroviruses of mice are incorporated directly into the genome as proviruses and are transmitted genetically from generation to generation. There is no effective treatment for leukemia, lymphoma, and mammary tumors in mice. Most mammary tumors in mice are adenocarcinomas [19]. Mammary tumors can be surgically excised but are likely to recur with a poor prognosis.

Other viral infections in mice

Mice have been reported to seroconvert to a number of other viruses; however, these infections rarely result in the development of clinical disease. These include mouse parvovirus, mice minute virus, mouse poliovirus (Theiler's murine encephalomyelitis virus), pneumonia virus of mice, reovirus, adenovirus, polyoma virus, and cytomegalovirus.

Viral infections of guinea pigs

Respiratory disease in guinea pigs

Although there are viral causes of respiratory disease in guinea pigs, infections from other causes should be ruled out first, because they occur more frequently. For example, respiratory disease due to a bacterial infection is more common and may indicate an underlying stressor such as poor husbandry or nutritional management.

Adenovirus-induced respiratory disease with low morbidity and high mortality has been documented in guinea pigs [20–24]. Clinical signs range from those of pneumonia, including dyspnea, tachypnea, nasal discharge, rough hair coat, weight loss, and a hunched posture to acute death with no clinical signs. Postmortem evaluation of guinea pigs that die may reveal pulmonary consolidation and emphysema with petechiae, while histologic examination may reveal a necrotizing bronchitis and bronchiolitis with basophilic intranuclear inclusion bodies. Currently, there is no commercially available serologic test for adenoviral infection in guinea pigs because the agent is difficult to grow in culture. However, a PCR assay has been described [25].

It should be noted that there are reports in the literature describing infection with adenoviruses in guinea pigs without any concurrent clinical signs [25,26]. In addition, in experimental infection of guinea pigs with adenovirus-infected crude suspensions of lung tissue, only newborn guinea pigs (2–3 days old) displayed any clinical signs [27], suggesting other factors such as altered immune status are needed to induce adenoviral respiratory disease.

Reproductive disease in guinea pigs

Natural infection with cytomegalovirus in guinea pigs is usually subclinical; overt disease is rare [28]. The true prevalence of infection is unknown and likely to be low. However, fetal death and sow mortality have been reported in association with cytomegaloviral infection [29]. Sows may be depressed at presentation and die 2 to 3 days later, and they may have late abortions, stillbirths, or fetal death shortly after birth. Other clinical signs may include weight loss and lymphadenopathy. The virus may be transmitted transplacentally or by way of infected saliva or urine and can persist in the guinea pig as a latent infection for years [30]. An immunocompromised state, pregnancy, or host factors may play a role in the severity of clinical signs [31].

Neoplastic disease in guinea pigs

Based on ultrastructural examination of affected tissues, type C retroviral infections have been associated with leukemia and lymphoma in the guinea pig. Clinical signs included lymphadenopathy and a white blood cell count from 25,000 to 250,000/mm^3 [32]. Lymphoblastic cells were the predominant type and in some animals, infiltrated the liver and spleen, causing enlargement. The disease is often fatal within 2 to 5 weeks. Chemotherapy may be attempted, but there are no published reports of efficacy. The actual role of viral infections in naturally occurring guinea pig leukemia remains unknown. Several cases of naturally occurring leukemia were described in 1980 [33]; however, the role of viral infections in these cases was not determined.

Other viral infections in guinea pigs

There are rare reports of other guinea pig viral infections. These are usually subclinical in natural infection, produce clinical signs only in experimental infection, or the cause–effect relationship is still undetermined.

Although these agents are mentioned briefly here, clinicians should use caution when considering them as potential causes for disease. For example, although poliovirus infection has been suggested to cause hind limb lameness in pet shop guinea pigs, the clinical signs resolved with vitamin C treatment, thus complicating the association of poliovirus and disease [34]. Seroconversion to paramyxoviruses (caviid parainfluenza virus 3, Sendai virus, pneumonia virus of mice, simian virus 5) has been described in guinea pigs. No clinical signs have been seen with natural infections; however, histologic evidence of mild pneumonia has been identified in infected guinea pigs [35,36]. There is one report of a coronavirus-like particle causing a wasting syndrome in 3- to 4-week-old guinea pigs in which no other cause of disease could be identified [37]. Clinical signs included anorexia, weight loss, diarrhea, and death.

Viral infections of Syrian hamsters

Like other rodents, spontaneous viral infections that result in clinical signs are uncommon in the hamster. A discussion of hamster viral infections also must include lymphocytic choriomeningitis virus (LCMV). The latter is most important because of its zoonotic disease causing potential.

Systemic disease in hamsters

LCMV is an RNA virus of the arenavirus group. Natural infections are uncommon and asymptomatic; however, this virus can infect humans, and many reported human cases have been associated with infected hamsters, including pets (see the review by Parker [38]). This may be because infected hamsters shed larger amounts of virus in their urine and may develop a more persistent viruria when compared with other rodents [38,39].

The manifestation of disease in hamsters experimentally infected with LCMV may give some insight into natural disease progression [39]. Disease manifestation varies with virus and hamster strain and the age of inoculation [40,41]. Experimental in utero or perinatal infections produce a subclinical persistent infection or a chronic, progressive wasting disease associated with immune complex glomerulonephritis, vasculitis, and multi-organ inflammation. Approximately half of these infected hamsters cleared the infection. Of note, similar signs were seen in pups born to viremic mothers, which conceivably could occur in a natural setting. When infected as young adults, hamsters developed viruria and viremia for up to 6 months, but did not exhibit clinical signs [39].

Because of the zoonotic potential of LCMV, screening of breeding colonies supplying the pet trade may be warranted. Diagnosis of asymptomatic infections is based on detection of serum antibody to LCMV by way of ELISA. PCR analysis of tissues can also be used to confirm persistent or acute infections [42]. All hamsters that are infected with LCMV or are at risk for infection should be euthanatized, and their cages and environment should be thoroughly disinfected.

Neoplastic disease in hamsters

Hamster polyomavirus is a DNA oncogenic virus that induces two neoplastic syndromes in hamsters depending on the age of the animal at the time of infection. Young hamsters develop a multicentric lymphoma involving mesenteric lymph nodes and abdominal viscera [43]. The other neoplastic syndrome occurs in adult hamsters and in endemically infected colonies, where viral infection results in trichoepitheliomas (Fig. 3) [30]. A presumptive diagnosis can be made on gross findings of epitheliomas or lymphomas. Confirmation requires histopathologic examination or PCR analysis of neoplastic tissue or kidney [43].

Musculoskeletal disease in hamsters

Recently, a case of naturally occurring spontaneous disease due to hamster parvovirus infection was described [44]. The disease manifested as decreased litter sizes with clinical signs in 2- to 4-week-old hamsters. Signs included dome-shaped crania, potbellied appearance, testicular atrophy and discoloration, malformation, and loss of incisor teeth. These signs were reproduced with experimental infection studies, and higher-dose experimental infections yielded a multisystemic hemorrhagic disease. A presumptive diagnosis may be made based on clinical signs and gross lesions.

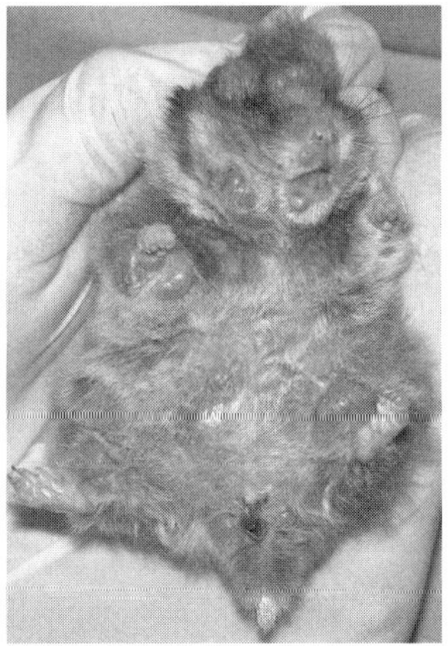

Fig. 3. Photograph of adult hamster infected with hamster polyomavirus demonstrating multiple trichoepitheliomas. (Courtesy of Dr. Joe Simmons, DVM, PhD, West Point, Pennsylvania).

Confirmation requires identification of parvovirus through PCR analysis [44,45].

Respiratory disease in hamsters

Sendai virus and pneumonia virus of mice are serologically distinct RNA viruses of the paramyxoviridae family. Incidence of seroconversion to these viruses is high. However, clinical disease and gross and histologic lesions have not been reported in naturally occurring infections, although rhinitis, bronchitis, and neonatal deaths have been reported with experimental infections [46,47].

Hamsters may also seroconvert to a number of viral pathogens of rodents. These include reovirus-3 and simian virus 5 [30,38]. The significance of these findings is poorly understood and clinical signs associated with seroconversion have not been reported. Adenoviral [48] and cytomegaloviral [49] inclusions have also been identified in hamsters, but to date, have not been associated with clinical disease.

Viral infections of chinchillas

Neurologic disease in chinchillas

Only two cases of naturally occurring viral infection in the chinchilla have been reported in the literature [50,51]; both are thought to be due to herpes or a herpes-like virus. The first report describes an adult female chinchilla that presented with a history of acute death. Necrotic foci with inclusion bodies were found in the spleen, pineal body, adrenal glands, and medulla of the brain. Electron microscopy was highly suggestive of a herpes-like viral infection based on morphology of the virus and distribution of lesions. In the second case, a 1-year-old male chinchilla with conjunctivitis, uveitis, and bilateral mydriasis for 2 weeks began to show neurologic signs, including disorientation, seizures, and biting at the cage bars, leading to atonic recumbancy. The chinchilla was treated with antibiotics and electrolytes but was euthanized after 3 weeks because of a poor prognosis. Significant pathologic findings included a unilateral purulent rhinitis, keratouveitis, retinopathy, optical neuritis, nonsuppurative meningitis, and polioencephalitis with neuronal necrosis. Intranuclear inclusion bodies were found in many areas of the brain and the nasal cavity. Electron microscopy, immunohistochemical staining, virus isolation, and nucleic acid sequence analysis confirmed an infection with human herpes simplex virus type 1, a common virus carried by humans causing oral or labial lesions that can lead to encephalitis. The authors proposed a primary ocular infection in the chinchilla potentially from contact to a human with a herpetic lesion that subsequently spread to the central nervous system. Although this route of transmission could not be confirmed, this raises the question of whether chinchillas can serve as reservoirs for herpes simplex virus.

Viral infections of prairie dogs

Systemic disease in prairie dogs

Until recently, there were no reports in the literature describing naturally occurring viral infections in prairie dogs. However, in May and June of 2003, an outbreak of monkeypox infection that spread from infected prairie dogs to humans occurred for the first time in the United States affecting multiple Midwestern states. Prairie dogs were exposed to monkeypox when housed in the same pet store as an infected Gambian giant rat and dormice [52] that were imported from Ghana. Animal-to-animal transmission of monkeypox was likely from respiratory and mucocutaneous routes [53]. Clinical signs of monkeypox infection in prairie dogs include anorexia, wasting, sneezing, coughing, lymphadenopathy, conjunctivitis, ocular and nasal discharge, papular skin lesions, tongue ulceration, and death [53,54].

Monkeypox is a member of the orthopoxvirus group that includes smallpox virus (variola) and vaccinia virus (the virus used in the smallpox vaccine). This viral disease occurs mostly in Africa, and many animal species are capable of being infected, including captive African monkeys (in which the virus was first recognized). In humans, clinical signs of monkeypox are similar to those of smallpox but are usually milder and include vesiculopustular skin lesions, fever, headache, muscle aches, lymphadenopathy, and lethargy [54]. Transmission from animals to people can occur via bite wounds, scratches, or exposure to infected blood, body fluids, or skin lesions. In people, mortality rates are reported to be from 1% to 10%; however, no human deaths occurred in the United States outbreak. The smallpox vaccine may be protective against monkeypox. No further cases of monkeypox in humans have been reported since June 22, 2003 [54].

There are no specific treatments for prairie dogs infected with monkeypox. If a case of monkeypox is suspected in an animal, precautions should be taken to prevent potential human exposure, including prior communication with the veterinary clinic before admitting the animal. Owners should isolate the potentially infected animal and wash hands, clothing, and any materials that have come in contact with the animal with household detergents or bleach. State or local health departments should be notified, and sample submission for diagnostic testing must be coordinated with state health departments. The Centers for Disease Control and Prevention (CDC) recommend that veterinarians should not perform biopsies or necropsies of animals with suspected monkeypox infection. Animals with suspected or confirmed infection with monkeypox should be humanely euthanatized. The CDC and the US Food and Drug Administration currently prohibit any sale, interstate movement, or release into the environment of prairie dogs within the United States. More information and guidelines can be found on the CDC Web site (www.cdc.gov/ncidod/monkeypox/index.htm).

Viral infections of hedgehogs

Neoplastic disease in hedgehogs

Retroviruses have been linked to multicentric skeletal sarcomas in African pygmy hedgehogs [55]. Retroviruses are often incorporated directly into the genome as proviruses, rendering attempts for diagnosis impractical. There are no established protocols for treatment of neoplasia in hedgehogs.

Systemic disease in hedgehogs

There has been one report of herpes simplex virus infection in an African pygmy hedgehog that was undergoing corticosteroid therapy for a prolapsed intervertebral disc [56]. Clinical signs were confined to anorexia. On necropsy, the liver was stippled with punctate off-white foci.

Viral infections of gerbils and sugar gliders

No naturally occurring viral diseases have been reported in gerbils or sugar gliders to date.

Viral zoonoses of small mammals

LCMV can infect many species, including the small mammals discussed in this article. If LCMV infection is suspected in a small mammal, those in contact with the potentially infected animal should seek medical attention immediately. Humans may become infected with LCMV from bite wounds or exposure to contaminated feces or urine, and may experience flu-like symptoms or rarely nonsuppurative meningitis as a result (see www.cdc.gov/ncidod/dvrd/spb/mnpages/dispages/lcmv.htm for more information regarding LCMV infection in people). Hamsters are the most common source of human exposure, most likely because they maintain a more persistent viruria than other mammals. However, it must be stressed that natural infection in hamsters and other small mammals is uncommon, and most recent human LCMV exposures have been linked to exposure in a laboratory setting either while handling experimentally infected immunocompromised research mice or through bench accidents.

Several species of wild rodents serve as reservoirs for hantavirus infection and are therefore considered a potential zoonotic hazard to humans. However, the likelihood of pet rodents acquiring this infection is extremely low. Simmons and Riley [36] offer an excellent review of hantavirus infections in rodents.

Rabies has been documented in rodents and other small mammals, but these numbers account for less than 1% of reported rabies cases in the United States and involve only feral rodents (http://www.cdc.gov/ncidod/

dvrd/rabies/Epidemiology/Epidemiology.htm). To illustrate this fact, between the years 1985 and 1994, there were 4 reported rabies cases in mice and rats, and these numbers were far surpassed by reports of the disease in feral woodchucks, raccoons, and rabbits [57]. Transmission of rabies to small mammals is extremely unlikely, because captive-bred animals have minimal exposure to the disease; furthermore, in the wild, when a small mammal comes in contact with a rabies-infected predator, the small mammal is more likely to be consumed than to have the disease transmitted during the encounter. Thus, rabies should be considered of very minimal concern to owners of the small mammals covered in this report.

Summary

Most viral infections in small mammals are transient and rarely produce clinical signs. When clinical signs do appear, they are often of a multifactorial etiology such as respiratory infection with Sendai virus and the bacteria *M pulmonis* in rodents. Diagnosis is generally made based on clinical signs, while therapy involves treatment for concurrent bacterial infections and supportive care. Small mammals may carry zoonotic viruses such as LCMV, but natural infections are uncommon. Viral diseases are rare (or largely unknown) for hedgehogs, chinchillas, and prairie dogs, while no known naturally occurring, clinically relevant viral diseases exist for gerbils and sugar gliders. This article is intended to aid the clinician in identifying viral infections in small mammals and to help determine the significance each virus has during clinical disease.

Acknowledgments

We thank Dr. Cynthia Besch-Williford for her critical review of the manuscript.

Further readings

Baker DG. Natural pathogens of laboratory animals. Washington, DC: ASM Press; 2003.

Committee on infectious disease of mice and rats. Infectious diseases of mice and rats. Washington, DC: National Academy Press; 1991.

Fox JG, Anderson LC, Loew FM, Quimby FW. Laboratory animal medicine. 2nd ed. Amsterdam: Academic Press; 2002.

Harkness JE, Wagner JE. The biology and medicine of rabbit and rodents. Baltimore (MD): Lea & Febiger; 1995.

Percy DH, Barthold SW. Pathology of laboratory rodents & rabbits. 2nd ed. Ames (IA): Iowa State University Press; 2001.

Quesenberry KE, Carpenter JW. Ferrets, rabbits, and rodents, clinical medicine and surgery. 2nd ed. St. Louis (MO): Saunders; 2004.

References

[1] Percy DH, Barthold SW. Rat. Viral infections. In: Percy DH, Barthold SW, editors. Pathology of laboratory rodents & rabbits. 2nd ed. Ames (IA): Iowa State University Press; 2001. p. 108–20.

[2] Macy JD Jr, Weir EC, Barthold SW. Reproductive abnormalities associated with a coronavirus infection in rats. Lab Anim Sci 1996;46:129–32.

[3] Utsumi K, Maeda K, Yokota Y, et al. Reproductive disorders in female rats infected with sialodacryoadenitis virus. Jikken Dobutsu 1991;40:361–5.

[4] Utsumi K, Yokota Y, Ishikawa T, et al. Reproductive disorders in female SHR rats infected with sialodacryoadenitis virus. Adv Exp Med Biol 1990;276:525–32.

[5] Vonderfecht SL, Huber AC, Eiden J, et al. Infectious diarrhea of infant rats produced by a rotavirus-like agent. J Virol 1984;52:94–8.

[6] Kilham L, Margolis G. Spontaneous hepatitis and cerebellar "hypoplasia" in suckling rats due to congenital infections with rat virus. Am J Pathol 1966;49:457–75.

[7] Coleman GL, Jacoby RO, Bhatt PN, et al. Naturally occurring lethal parvovirus infection of juvenile and young-adult rats. Vet Pathol 1983;20:49–56.

[8] Percy DH, Barthold SW. Mouse/viral infections. In: Percy DH, Barthold SW, editors. Pathology of laboratory rodents & rabbits. 2nd ed. Ames (IA): Iowa State University Press; 2001. p. 23–5.

[9] Committee on Infectious Disease of Mice and Rats. Lindsey JR, Boorman GA, Collin MJ Jr, et al. (Multiple systems). In: Infectious diseases of mice and rats. Washington, DC: National Academy Press; 1991. p. 236–9.

[10] Fenner F. Mousepox (infectious ectromelia): past, present, and future. Lab Anim Sci 1981; 31:553–9.

[11] Homberger FR, Zhang L, Barthold SW. Prevalence of enterotropic and polytropic mouse hepatitis virus in enzootically infected mouse colonies. Lab Anim Sci 1998;48:50–4.

[12] Committee on Infectious Disease of Mice and Rats. Lindsey JR, Boorman GA, Collin MJ Jr, et al. (Digestive system). In: Infectious diseases of mice and rats. Washington, DC: National Academy Press; 1991. p. 102–17.

[13] Cheever FS, Mueller JH. Epidemic diarrheal disease of suckling mice. III. The effect of strain, litter, and season upon the incidence of disease. J Exp Med 1948;88:309–16.

[14] Cheever FS, Mueller JH. Epidemic diarrheal disease of suckling mice. I. Manifestations, epidemiology, and attempts to transmit the disease. J Exp Med 1947;85:405–16.

[15] Zurcher C, Burek JD, Van Nunen MC, et al. A naturally occurring epizootic caused by Sendai virus in breeding and aging rodent colonies. I. Infection in the mouse. Lab Anim Sci 1977;27:955–62.

[16] Bhatt PN, Jonas AM. An epizootic of Sendai infection with mortality in a barrier-maintained mouse colony. Am J Epidemiol 1974;100:222–9.

[17] Parker JC, Reynolds RK. Natural history of Sendai virus infection in mice. Am J Epidemiol 1968;88:112–25.

[18] Itoh T, Kagiyama N, Iwai H, et al. Sendai virus infection in a small mouse breeding colony. Nippon Juigaku Zasshi 1978;40:615–8.

[19] Donnelly TM. Disease problems of small rodents. In: Quesenberry KE, Carpenter JW, editors. Ferrets, rabbits, and rodents: clinical medicine and surgery. 2nd ed. St.Louis (MO): Saunders; 2004. p. 299–315.

[20] Brennecke LH, Dreier TM, Stokes WS. Naturally occurring virus-associated respiratory disease in two guinea pigs. Vet Pathol 1983;20:488–91.

[21] Naumann S, Kunstyr I, Langer I, et al. Lethal pneumonia in guinea pigs associated with a virus. Lab Anim 1981;15:235–42.

[22] Harris IE, Portas BH, Goydich W. Adenoviral bronchopneumonia of guinea pigs. Aust Vet J 1985;62:317.

[23] Feldman SH, Richardson JA, Clubb FJ Jr. Necrotizing viral bronchopneumonia in guinea pigs. Lab Anim Sci 1990;40:82–3.

[24] Eckhoff G, Mann P, Gaillard E, et al. Naturally developing virus-induced lethal pneumonia in two guinea pigs (*Cavia porcellus*). Contemp Top Lab Anim Sci 1998;37:54–7.

[25] Butz N, Ossent P, Homberger FR. Pathogenesis of guinea pig adenovirus infection. Lab Anim Sci 1999;49:600–4.

[26] Crippa L, Giusti AM, Sironi G, et al. Asymptomatic adenoviral respiratory tract infection in guinea pigs. Lab Anim Sci 1997;47:197–9.

[27] Kunstyr I, Maess J, Naumann S, et al. Adenovirus pneumonia in guinea pigs: an experimental reproduction of the disease. Lab Anim 1984;18:55–60.

[28] Van Hoosier GL Jr, Giddens WE Jr, Gillett CS, et al. Disseminated cytomegalovirus disease in the guinea pig. Lab Anim Sci 1985;35:81–4.

[29] Motzel SL, Wagner JE. Diagnostic exercise: fetal death in guinea pigs. Lab Anim Sci 1989; 39:342–4.

[30] Percy DH, Barthold SW. Guinea pig/viral infections. In: Percy DH, Barthold SW, editors. Pathology of laboratory rodents & rabbits. 2nd ed. Ames (IA): Iowa State University Press; 2001. p. 213–6.

[31] Harkness JE, Murray KA, Wagner JE. Biology and diseases of guinea pigs. In: Fox JG, Anderson LC, Loew FM, editors. Laboratory animal medicine. 2nd ed. Amsterdam: Academic Press; 2002. p. 203–47.

[32] Van Hoosier GL Jr, Robinette LR. Viral and chlamydial diseases. In: Wagner JE, Manning PJ, editors. The biology of the guinea pig. New York: Academic Press; 1976. p. 137–52.

[33] Hong CC, Liu PI, Poon KC. Naturally occurring lymphoblastic leukemia in guinea pigs. Lab Anim Sci 1980;30:222–6.

[34] Hansen AK, Thomsen P, Jensen HJ. A serological indication of the existence of a guinea pig poliovirus. Lab Anim 1997;31:212–8.

[35] Blomqvist GA, Martin K, Morein B. Transmission pattern of parainfluenza 3 virus in guinea pig breeding herds. Contemp Top Lab Anim Sci 2002;41:53–7.

[36] Simmons JH, Riley LK. Hantaviruses: an overview. Comp Med 2002;52:97–110.

[37] Jaax GP, Jaax NK, Petrali JP, et al. Coronavirus-like virions associated with a wasting syndrome in guinea pigs. Lab Anim Sci 1990;40:375–8.

[38] Parker JC. Viral diseases. In: Van Hoosier GLJ, McPherson CW, editors. Laboratory hamsters. Orlando (FL): Academic Press; 1987. p. 95–111.

[39] Parker JC, Igel HJ, Reynolds RK, et al. Lymphocytic choriomeningitis virus infection in fetal, newborn, and young adult Syrian hamsters (Mesocricetus auratus). Infect Immun 1976;13:967–81.

[40] Genovesi EV, Peters CJ. Susceptibility of inbred Syrian golden hamsters (Mesocricetus auratus) to lethal disease by lymphocytic choriomeningitis virus. Proc Soc Exp Biol Med 1987;185:250–61.

[41] Genovesi EV, Johnson AJ, Peters CJ. Susceptibility and resistance of inbred strains of Syrian golden hamsters (Mesocricetus auratus) to wasting disease caused by lymphocytic choriomeningitis virus: pathogenesis of lethal and non-lethal infections. J Gen Virol 1988; 69:2209–20.

[42] Besselsen DG, Wagner AM, Loganbill JK. Detection of lymphocytic choriomeningitis virus by use of fluorogenic nuclease reverse transcriptase-polymerase chain reaction analysis. Comp Med 2003;53:65–9.

[43] Simmons JH, Riley LK, Franklin CL, et al. Hamster polyomavirus infection in a pet Syrian hamster (Mesocricetus auratus). Vet Pathol 2001;38:441–6.

[44] Besselsen DG, Gibson SV, Besch-Williford CL, et al. Natural and experimentally induced infection of Syrian hamsters with a newly recognized parvovirus. Lab Anim Sci 1999;49: 308–12.

[45] Besselsen DG, Besch-Williford CL, Pintel DJ, et al. Detection of newly recognized rodent parvoviruses by PCR. J Clin Microbiol 1995;33:2859–63.

[46] Profeta ML, Lief FS, Plotkin SA. Enzootic sendai infection in laboratory hamsters. Am J Epidemiol 1969;89:316–24.

[47] Percy DH, Barthold SW. Hamster/viral infections. In: Percy DH, Barthold SW, editors. Pathology of laboratory rodents & rabbits. 2nd ed. Ames (IA): Iowa State University Press; 2001. p. 169–74.

[48] Gibson SV, Rottinghaus AA, Wagner JE, et al. Naturally acquired enteric adenovirus infection in Syrian hamsters (Mesocricetus auratus). Am J Vet Res 1990;51:143–7.

[49] Lussier G. Murine cytomegalovirus (MCMV). Adv Vet Sci Comp Med 1975;19:223–47.

[50] Wohlsein P, Thiele A, Fehr M, et al. Spontaneous human herpes virus type 1 infection in a chinchilla (Chinchilla laniger f. dom.). Acta Neuropathol (Berl) 2002;104:674–8.

[51] Goudas P, Giltoy JS. Spontaneous herpes-like viral infection in a chinchilla (Chinchilla laniger). Wildl Dis 1970;6:175–9.

[52] Centers for Disease Control and Prevention. Update: multistate outbreak of monkeypox— Illinois, Indiana, Kansas, Missouri, Ohio, and Wisconsin, 2003. MMWR 2003;52:642–6.

[53] Guarner J, Johnson BJ, Paddock CD, et al. Monkeypox transmission and pathogenesis in prairie dogs. Emerg Infect Dis 2004;10:426–31.

[54] Reed KD, Melski JW, Graham MB, et al. The detection of monkeypox in humans in the Western Hemisphere. N Engl J Med 2004;350:342–50.

[55] Peauroi JR, Lowenstine LJ, Munn RJ, et al. Multicentric skeletal sarcomas associated with probable retrovirus particles in two African hedgehogs (*Atelerix albiventris*). Vet Pathol 1994;31:481–4.

[56] Allison N, Chang TC, Steele KE, Hilliard JK. Fatal herpes simplex infection in a pygmy African hedgehog (Atelerix albiventris). J Comp Pathol 2002;126:76–8.

[57] Childs JE, Colby L, Krebs JW, et al. Surveillance and spatiotemporal associations of rabies in rodents and lagomorphs in the United States, 1985–1994. J Wildl Dis 1997;33:20–7.

ELSEVIER
SAUNDERS

Vet Clin Exot Anim 8 (2005) 123–138

VETERINARY
CLINICS
Exotic Animal Practice

Viral diseases of the rabbit

Aric P. Krogstad, DVM*, Janet E. Simpson, DVM,
Scott W. Korte, DVM

*Department of Veterinary Pathobiology, College of Veterinary Medicine, University of
Missouri-Columbia, W108 Veterinary Medical Building 1600 East Rollins,
Columbia, MO 65211, USA*

This article is designed to be a practical reference for veterinarians who see rabbits in private practice. We have not attempted to thoroughly cover major aspects of husbandry, biology, or nonviral disease; the reader is referred to other references for these purposes [1–4]. We have emphasized naturally-occurring viral infections that cause clinically significant disease. This emphasis comes at the expense of discussing viral pathology and alterations in physiology that may be important in research settings but have less clinical impact in a practice setting.

This paper is subdivided into sections based on the organ system that is affected predominantly by infection. We recognize that this format may oversimplify the multi-systemic effects of viral diseases but believe that the approach will be helpful for easy reference and discussion of the most common clinical presentations. Each section includes a brief overview of the viral agent, host susceptibility, disease transmission, clinical signs, diagnosis, control, and prognosis. Important differential diagnoses for disease presentation also are described when appropriate.

Treatment of viral disease centers on supportive care; in some cases, the prognosis may be grave. We recommend that practitioners make use of their clinical judgment when dealing with a sick rabbit, including aggressive treatment (eg, encouraging gastric motility), meeting the patient's fluid needs, and removing nonessential stressors. In a multiple rabbit household or in a rabbitry, control measures to minimize the contagious nature of viral disease must be implemented. In rare instances, it may be appropriate to discuss the possibility of zoonotic disease transmission.

This work was supported in part by NIH grant T32RR07004 and Mission Enhancement Funds from the University of Missouri.

* Corresponding author.

E-mail address: krogstada@missouri.edu (A.P. Krogstad).

doi:10.1016/j.cvex.2004.09.002   vetexotic.theclinics.com

Viral diseases of the neurologic system

Viral diseases that affect the central nervous system of rabbits rarely have been reported; however, rabies and herpes simplex infections have been associated with neurologic disease in this species [5–7].

Rabies

Between 2000 and 2002, eight laboratory-confirmed cases of rabies in rabbits (*Oryctolagus cuniculus*) were reported to the Centers for Disease Control and Prevention. All reports occurred in states in which rabies is enzootic in raccoons [8–10]. In the few clinical case reports that documented rabies infection of the domestic rabbit (*O cuniculus*), a paralytic form of the disease was observed principally. In these cases, infected rabbits had a history of being housed in an outdoor situation and had confirmed or potential exposure to wildlife [5,11–13]. Initial signs of disease in infected rabbits have been nonspecific, including anorexia and lethargy; in one case, these were the only signs that were observed [12]. As the disease progresses, neurologic signs, including head tremors, blindness, and ascending paralysis, have been described; the affected rabbit often dies within 3 or 4 days [5,11–13].

Rabies is diagnosed in animals by means of postmortem direct immunofluorescent antibody testing of brain tissue by state or local health departments [14]. Although rabies is rare in rabbits, they are susceptible to infection and initial signs of infection may be nonspecific in nature. Rabies infection should be considered as a rule-out in rabbits who demonstrate symptoms of a rapidly progressing neurologic disease with a history of outdoor housing or possible exposure to wildlife or other potentially rabid animals. No approved rabies vaccination for use in rabbits is marketed in the United States [5]. Clients who keep rabbits outdoors should be cautioned to protect their animals from potential exposure to feral animals.

Herpes virus hominus

Naturally-occurring disease that is associated with herpes virus hominus infection has been reported in the domestic rabbit; there have been two reports of clinical disease associated with this virus. In both cases, a person who had a herpetic lesion (eg, cold sore) was known to have close contact with the infected animals [6,7]. Infected rabbits exhibited signs of anorexia and neurologic abnormalities, including restlessness, circling, and tonic-clonic spasms. In both cases, neurologic disease progressed to lateral recumbency at which point the animals were euthanized. On histopathology, a nonsuppurative meningoencephalitis with neural cell necrosis and intra-nuclear inclusion bodies was observed. The agent involved in the infection was classified as herpes virus hominus subcategory herpes simplex virus by polymerase chain reaction (PCR) and immunohistochemistry.

Viral diseases of the integument

Viral infections of the integument are more common in the rabbit than in most species. The viruses that are associated with skin disease in the rabbit belong to the pox and papova virus families.

Myxoma virus

Myxoma virus is a leporipoxvirus of the Poxviridae family [15]. In 1896, myxoma virus infection was documented in European rabbits that were located in Uruguay. This outbreak was associated with high mortality and numerous mucinous skin masses on affected animals [16]. In 1928, myxoma virus was first recognized in North America when an outbreak of fatal disease was observed in California [17].

Myxoma virus is transmitted principally by arthropod vectors, mosquitoes, and fleas. Rabbits of the genus *Sylvilagus* serve as the natural host for myxoma virus; in them, myxomatosis produces a benign skin disease that is associated with the development of local skin tumors that resemble fibromas at the site of viral inoculation by way of bite by mosquito or other blood-sucking insect [18]. In the European hare (*Lepus europaeus*), myxoma virus infection rarely has been associated with disease. Infection in the European or domesticated rabbit (*O cuniculus*) has been associated with systemic disease with mortality rates of up to 100% [19]; however, the presentation of disease depends greatly upon the strain of virus and the particular species of rabbit [20,21]. Typical skin lesions that are associated with disease in the domestic rabbit include edema of the eyelids, ears, nose, anus, and genitals; blepharoconjunctivitis; hemorrhage; and nodules on the ears, head, body, and legs at the site of infection that may become congested or necrotic [17,19,22–24]. Histologically, the skin lesions consist of the proliferation of stellate (myxoma) mesenchymal cells surrounded by a mucinous matrix. In addition, endothelial cell proliferation, intracytoplasmic inclusions in various cell types, and epidermal cell hyperplasia or degeneration in advanced lesions may be observed [19,22,24,25].

For a thorough description of the systemic disease that is associated with myxoma virus refer to the section that details viral multi-systemic diseases.

Rabbit (Shope) fibroma virus

Rabbit fibroma virus is a member of the Poxviridae family and is related closely antigenically to myxoma virus [26]. Rabbit fibroma virus was first identified in an Eastern cottontail rabbit (*S floridanus*) in 1932 [27]. This species was shown to serve as the natural host for the virus, but other species of cottontail rabbits and European rabbits (*O cuniculus*) have proven to be susceptible as well [19,27]. Initially, this virus was considered to be a benign skin disease of cottontail rabbits; however, it since has been associated with an epizootic of disease in domestic rabbits that resulted in high morbidity and mortality in newborn animals. Neonatal infection was associated with systemic disease that was characterized by lethargy, poor body condition, and death [28].

Rabbit fibroma virus infection has been recognized in cottontail rabbits in the United States and Canada [18]. The natural cycle of infection is not understood completely, but biting arthropods may serve as the primary means for virus transmission [29–31].

Infection in European rabbits (*O cuniculus*) is characterized by flat, freely movable, subcutaneous tumors that primarily are located along the legs, feet, ears, muzzle, and around the eyes (Fig. 1). These tumors may measure several centimeters in diameter and resolve spontaneously within a few months. Microscopic examination of the skin lesions reveal proliferation of mesenchymal cells which become stellate or ovoid, multifocal areas of necrosis, and mononuclear and polymorphonuclear infiltrates [28,32–34].

Diagnosis of rabbit fibroma virus is based upon clinical presentation and histopathologic examination of biopsied masses. Differentiation of rabbit fibroma virus from infection with myxomatosis and papillomatosis in the domestic rabbit should be accomplished readily based upon characteristic gross lesions [17]. Control of disease may be undertaken by limiting

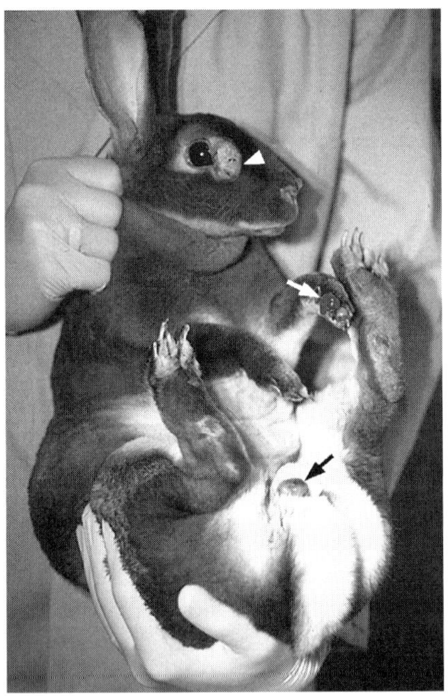

Fig. 1. Rabbit (Shope) fibroma virus. Photograph of a Rex rabbit that presented with classic multi-focal subcutaneous masses located around the eyes, digits, and in the genital region subsequent to rabbit fibroma viral infection. (Courtesy of Craig Franklin, DVM, PhD, DACLAM, University of Missouri, Columbia, Missouri).

exposure to arthropod vectors. Because tumors regress spontaneously, treatment, in general, is not necessary [35].

Cottontail rabbit (Shope) papillomavirus

The cottontail rabbit (Shope) papillomavirus was the first oncogenic virus to be identified in mammals. In 1933, this virus was found to be associated with the appearance of wartlike tumors in the cottontail rabbit (*S floridanus*) [36]. Although the cottontail rabbit serves as the natural host, this virus is transmissible to domestic rabbits (*O cuniculus*) and was responsible for a spontaneous outbreak of papillomatosis in domestic rabbits from southern California [37,38].

Cottontail papillomavirus has been associated with disease in the cottontail rabbit population in the midwestern United States and in domestic rabbits in California [38,39]. This virus may spread by direct contact; however, the principal mode of transmission occurs through arthropod vectors. The rabbit tick (*Haemaphysalis leporis-palustris*) is suspected to be the most common mode of transmission in the cottontail rabbit; however, experimental transmission of this virus also was demonstrated by mosquitoes and reduviid bugs [40,41]. In the domestic rabbit, the predominant location of papillomas along hairless areas of the ears and eyelids suggests that transmission from the cottontail rabbit to the domestic rabbit may occur primarily through the mosquito [38].

In the cottontail rabbit, infection is characterized by wartlike protuberances that are located predominantly along the neck, shoulders, and abdomen. At the sites of infection, these warts start as red, raised areas that become papillomas and eventually may develop into large, keratinized horny growths. These growths may vanish within a few months or they may become neoplastic in nature and be replaced by squamous cell carcinomas [36,42]. In the domestic rabbit, spontaneous infection was characterized by horny protuberances that were located most commonly along the ears and eyelids (Fig. 2) [38]. Experimental infection of the domestic rabbit revealed a lesser degree of papilloma regression than was observed with the cottontail rabbit and a high rate of carcinoma development following infection [43].

Cottontail papillomavirus may be diagnosed based upon characteristic clinical presentation and histopathologic examination. In areas where infection is endemic, natural infection may be controlled through control of the arthropod vector. Treatment may be accomplished through surgical removal of the papilloma [35].

Viral diseases of the gastrointestinal system

Gastrointestinal disease in rabbits is common and accounts for a significant economic loss. Viral diseases that affect the gastrointestinal system of rabbits include papillomavirus, parvovirus, rotavirus, and enteric

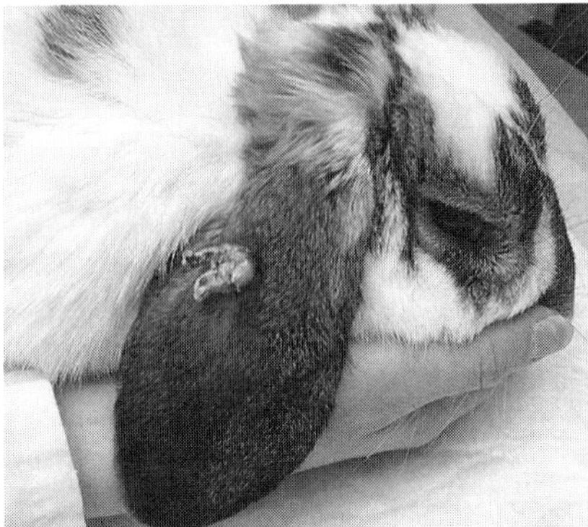

Fig. 2. Cottontail rabbit (Shope) papillomavirus, papillomatosis. Photograph of a domestic rabbit that presented with an approximately 1.5-cm keratinized horny protuberance located along the dorsal surface of the ear pinna secondary to cottontail rabbit papillomavirus. (Courtesy of Craig Franklin).

corona virus. Prominent nonviral causes of gastrointestinal disease include colibacillosis, coccidiosis, and enterotoxemia [4]. Coinfection by viral and nonviral pathogens is common and often cause more severe clinical signs than the viral pathogen alone would induce.

Papillomavirus

Oral papillomatosis in the rabbit is caused by a papillomavirus in the Papovaviridae family and is not related antigenically to the Shope papilloma virus [44]. The prevalence of oral papillomatosis in the rabbit has been described as "infrequent." The estimated prevalence has been reported to be between 5% and 33% [44–47]. Infection has been noted most frequently in animals between the ages of 2 months and 2 years [44,46,47].

Infection with oral papillomavirus is characterized by the development of small (1–2 mm) papillomas that usually occur on the ventral aspect of the tongue, although other locations on the tongue and in the oral cavity have been reported [46–48]. One report described lesions that were as large as 10 mm behind the mandibular incisors [45]. Gross lesions have been noted as soon as 14 days after experimental infection and reach their maximal size after approximately 1 month; most lesions begin to regress spontaneously [45,46,48]. Experimental infection only has been demonstrated when the rabbits were inoculated following artificial damage to the epithelium of the tongue and oral cavity [48]. This suggests that rough or hard food, chewing on rough cage bars, or malocclusion may predispose animals to infection with the oral papillomavirus. Animals that were exposed previously to oral papillomavirus (seropositive) were resistant to reinfection [46]. The duration of this immunity is not known.

Microscopically, oral papillomavirus lesions consist of proliferating epithelial cells with hyperkeratosis, with or without a fibrovascular stalk.

Cells at the junction of the stratum spongiosum and the papilloma have abundant, clear cytoplasm with eccentric nuclei and basophilic intranuclear inclusions [44,46,48].

Affected animals show few clinical signs and the disease usually is diagnosed incidentally upon finding the characteristic papillomas on the tongue or in the oral cavity during preparation for surgery or during necropsy [45,47]. The absence of obvious clinical signs may account, in part, for the variation in reported prevalence.

There are few differential diagnoses for oral papillomas in the rabbit; however, one case report noted a sialocele (ranula) on necropsy that was mistaken for an oral papillomavirus lesion [49]. The diagnosis of rabbit oral papillomatosis usually is based upon demonstration of the characteristic papillomatous lesion in the oral cavity. Because the lesions are transient and clinical signs usually are not associated with the disease, treatment or removal of oral papillomas is unnecessary in most cases. Although natural transmission to other species has not been noted, hamsters developed fibromas following experimental infection with rabbit oral papilloma virus [46].

Parvovirus

Just as in many other species, parvovirus infection in rabbits was shown to produce clinical disease [50,51]. Lapine parvovirus (LPV) was first isolated from rabbits in Japan in 1977. This investigation found antibodies to LPV in 46.7% of 90 commercial rabbits that were tested [52]. A similar study that was conducted in the United States using rabbits (*O cuniculus*) that were held in a laboratory setting demonstrated that 75% of 46 animals were seropositive for LPV [51].

Although the percentage of animals that is seropositive is high, the clinical signs that are caused by the disease are mild. Following experimental inoculation, clinical signs included listlessness and anorexia 4 to 6 days postinoculation. Virus was recovered from the liver, pancreas, spleen, small intestine, cecum, mesenteric lymph nodes, and feces for 14 days following inoculation. Microscopically, the virus caused mild to moderate catarrhal enteritis with hyperemia, exfoliation of enterocytes, and luminal fluid accumulation [50].

Given the short duration and minimal clinical signs, diagnosis of parvoviral disease in rabbits may prove to be difficult in a clinical setting. Isolation or demonstration of viral particles from the feces is the primary diagnostic method. To the authors' knowledge, a serologic test is not commercially available. Treatment of uncomplicated parvoviral infection is supportive.

Rotavirus

Rotaviral infections were shown to cause diarrhea in a wide variety of host species. Rotaviruses are a member of the Reoviridae family. Initially, rotavirus was isolated from rabbits that had diarrhea in England in 1976 [53]. Since that time, rabbit rotavirus has been studied in depth and has been used in modeling the immune response and in developing vaccines for human rotaviral infection [54]. Studies showed that the rotaviral strains that infect rabbits are group A serotype 3, subgroup I and II rotaviruses [55,56]. These serotypes are common among a variety of species, including humans and cattle [56,57].

Although the true incidence of rotaviral disease in rabbits is not known, antibodies were demonstrated in 29% of 17 wild cottontail rabbits (*S floridanus*) in Ontario, 52% of 27 snowshoe hares (*L americanus*) in the Yukon, and 98% of 91 commercially-raised New Zealand White rabbits (*O cuniculus*) from two rabbitries in Ontario [58]. The presence of rotaviral antibody in rabbits was shown to be age-dependent. More than 90% of neonates (<1 month), weanlings (2–3 months), young adults (3–4 months), and breeding adults (>5 months) had antibodies to rotavirus, whereas only 25% to 69% of preweanling (1–2 months) animals had antibodies to rotavirus [59,60]. This corresponds with other reports that show the peak prevalences of rotaviral disease in endemically-infected rabbitries is in animals that are between 36 and 42 days old [61]. This suggests that animals become infected after maternal antibodies wane. In naïve and experimentally-inoculated colonies, 1- to 2-week-old animals were the most severely affected [62,63]. Following experimental infection, animals between 1 and 2 weeks old were the only age group that showed clinical signs. Fecal viral shedding is similar for all age groups and tapers around 10 days after experimental infection [63].

The most common clinical signs in rabbits that have rotaviral disease are diarrhea and dehydration [53,62,64,65]; however, rotaviral infections may be clinically silent, with the affected animals shedding virus in their feces [53,54,60,66]. Experimental infections with rotavirus that was isolated from the feces of animals that had severe diarrhea resulted in only mild clinical signs [63–65]. This suggests that other factors may be involved when rotavirus is isolated in outbreaks of severe diarrhea and a high mortality rate. This is supported by a study that demonstrated more severe diarrhea and an increased mortality rate when animals were coinfected with *Escherichia coli* [65]. On necropsy, animals that are infected with rotavirus have a distended, fluid-filled small intestine and cecum [58,64]. Histopathologic findings spanned the jejunum and the ileum; lesions included shortened, fused villi with attenuated epithelium and slight to moderate increases in crypt depth [62,64].

A clinical diagnosis may be made based upon clinical signs in the proper age group or viral isolation from the feces of animals that have clinical signs. An antigen capture ELISA, such as those used in rotaviral diagnosis in cattle [67], also may be useful given the similarity in serotypes. Differential diagnoses include coccidiosis, Tyzzer's disease (*Clostridium piliforme*), clostridial enteritis/toxemia, colibacillosis (*E coli*), and cecal dysbiosis (mucoid enteropathy) [4].

Treatment of an individual animal relies upon fluid and electrolyte replacement, either orally or subcutaneously. Treatment in a rabbitry might include discontinuing breeding for 4 to 6 weeks to stop the influx of naïve animals and to allow infected animals adequate time to mount an immune response and stop shedding the virus [18]. After an individual animal has recovered, it is resistant to subsequent infection [54,66,67].

Coronavirus

Coronaviral infections in rabbits have two manifestations: (1) enteritis and (2) pleural effusion disease and cardiomyopathy [68]. Cardiomyopathy was reported following coronaviral experimental infection in a laboratory setting; however, there are no reports to suggest that this occurs as sequelae to naturally-occurring infection [2,69–71]. Coronavirus was linked first to diarrhea in rabbits in Canada in 1980 when viral particles that were consistent with the Coronaviridae family were noted on electron microscopic evaluation of fecal matter from rabbits that presented with diarrhea [72]. In a survey of rabbitries from the northwestern United States and southwestern Canada, the prevalence of antibodies to rabbit enteric coronavirus (using a canine coronavirus test) ranged from 3% to 33%. Rabbitries with a high prevalence of diarrhea were more likely to have antibody titers to rabbit enteric coronavirus (RECV). The cross-reactivity with canine coronavirus (type I coronavirus) and lack of cross-reactivity with mouse hepatitis virus (type II coronavirus) suggest that RECV is a type I coronavirus [73]. Viral particles have been isolated from the feces of experimentally-infected rabbits as long as 29 days postinfection [74].

Clinical signs of RECV infection in rabbits include watery diarrhea, abdominal distention, anorexia, and sudden death [68,72,74]. Rabbits that are between 3 and 10 weeks old are affected most commonly [72]. Experimentally infected animals have not shown sudden death as a clinical sign; this suggests that like rotaviral infections, concurrent infections with other organisms may exacerbate clinical signs. Experimental infection has revealed that gross lesions (eg, congested small intestines and fluid cecal contents) may be noted 6 hours to 3 days postinoculation. On histopathologic examination of the small intestine, necrosis of the villous epithelial cells has been noted within 6 hours of experimental infection. By 48 hours postinoculation, small intestinal villous blunting, crypt hypertrophy, complete M-cell necrosis, and necrosis of villous epithelial cells overlying the gut-associated lymphoid tissue was noted. In addition, RECV may result in subclinical infections [74].

Diagnosis may be made by finding viral particles in the feces of rabbits that have diarrhea or characteristic lesions on necropsy and histologic evaluation of the gastrointestinal tract. A commercially-available serologic assay for RECV is not known to the authors but it may be possible to detect RECV by using a commercial canine coronavirus test [67]. In addition, use of serology may be of limited immediate utility since animals that develop diarrhea are unlikely to have mounted an antibody response to RECV.

Treatment of individual animals includes isolation from unaffected rabbits and supportive therapy. Management changes in the face of an outbreak (eg, delayed weaning, diet changes, broad spectrum antibiotic treatment, vaccination) have not been successful in preventing morbidity and mortality [75].

Viral multi-systemic diseases

Rabbit caliciviral disease

Rabbit viral hemorrhagic disease (RHD) is related closely to the European brown hare syndrome (EBHS); both diseases are caused by related, but antigenetically distinct, caliciviruses [2,76,77]. RHD is an extremely contagious disease and is the only rabbit disease that is reportable to the U.S. Department of Agriculture. Disease was first reported in China in 1984 and is now endemic in New Zealand and Australia. Outbreaks of disease in American domestic rabbits were documented recently [78,79]. A recent outbreak in Illinois resulted in the euthanasia of more than 4800 rabbits to stop the spread of disease [78]. If RHD is suspected, the state and federal veterinarian should be contacted immediately and quarantine should be instituted.

Confirmation of disease can be performed with PCR of tissue extracts, hemagglutination assay, and ELISA; however, histopathologic examination of the liver and spleen also may serve as a means of diagnosis [78,80]. Clinical signs may include nonspecific illness, fever, convulsions, morbidity of more than 30%, and mortality of 90% or more in affected domestic (*O cuniculus*) rabbit populations. Disseminated intravascular coagulopathy plays a prominent role in disease pathology; gross findings at necropsy may include blood-stained nasal discharge, hepatomegaly, splenomegaly, and hemorrhage on surfaces, including the pericardium and intestine [2,81]. EBHS has been reported to affect wild and farmed hares (*L timidus* and *L europaeus*) in many European countries; similar to RHD, outbreaks of explosive disease are possible with similar clinical signs and prevalences of mortality in hares that are affected with EBHS [82,83]. Experimental infection of domestic rabbits with EBHS calicivirus did not result in clinical disease and did not protect animals from subsequent challenge with RHD virus [84]. Outbreaks of RHD in domestic rabbits may occur rapidly owing to an incubation time of only 1 to 2 days and death in 2 to 3 days; fomites, direct contact, aerosols, insect vectors, and carcasses serve as means of spreading disease [2,85].

Herpesvirus

Sudden death in Canadian rabbitries (*O cuniculus*) has been attributed to a herpeslike viral infection. Antemortem signs of disease may be subtle or absent but gross necropsy findings include hydropericardium and multifocal hemorrhages in the skin, abdominal viscera, and heart. Although the prevalence and pathogenesis of this virus is unclear, histopathologic lesions, including intranuclear inclusions that are consistent with herpes virus and isolation of herpesviral particles from lesioned tissues, were found. These findings led some investigators to conclude that herpeslike viral infection was responsible for the outbreak [2].

In wild populations of rabbits, including the eastern cottontail (*S floridans*), serologic surveys demonstrated a low (<4%) prevalence of serum antibodies to *Herpesvirus sylvilagus* [86]. The clinical significance of this is unclear; however, in rare cases, unexplained death in cottontails may be attributed to herpesvirus [87]. Experimental inoculation of pregnant cottontail rabbits demonstrated that this virus is not transmitted across the placenta [88]. Recent studies demonstrated that experimental inoculation of cottontails with *H sylvilagus* resulted in the development of lymphoproliferative lesions that were similar to lesions noted following Epstein-Barr virus infection in humans [89,90].

Myxomatosis

Myxoma virus in its natural hosts (*S brasiliensis* or *S bachmani*) causes only a benign fibroma. In the European or domestic rabbit (*O cuniculus*), however, it can cause a fulminant disease that is known as myxomatosis (or the amyxomatous form of myxomatosis). After its initial introduction into European rabbit populations in Europe and Australia, the virus killed nearly 100% of infected animals [91]; however, the combination of selection for resistant hosts and attenuated strains of virus has greatly reduced the prevalence of mortality due to disease. A study that was performed in Spain demonstrated endemic myxomatosis in wild European rabbit populations, but found no evidence of adult mortality that was associated with this prevalence [92]. A recent study in which naïve European rabbits were infected experimentally with several strains of myxoma virus provided additional evidence of the variable virulence of different strains [93]. Experimental infection of European rabbits with an attenuated form of myxoma virus resulted in transient systemic illness with orchitis and epididymitis; animals recovered from infection and demonstrated normal fertility by 60 to 90 days following infection [94]. There are occasional outbreaks of disease in domestic rabbits in the United States (most commonly California) and Mexico which are believed to represent accidental transmission from the brush rabbit (*S brushmani*) by way of an insect vector [95].

Prognosis and presence or absence of clinical signs vary tremendously with host and viral strain. Clinical signs may be consistent with nonspecific illness and hypothermia (unlike the fever associated with RHD); gross pathology of dead animals may range from edema of the eyelids, conjunctiva, and anogenital region to extensive lung lesions (associated with secondary bacterial infections) [2,25,93]. Diagnosis may be performed through PCR of tissue extracts, gross or microscopic lesions, or inoculation of suspected infected tissue into susceptible rabbits [2]. Treatment consists of supportive care and prognosis varies significantly with strain and host. Vaccine development is on-going but no vaccine is marketed in the United States [96,97]. Clients in California or other areas with increased risk of disease, should be counseled to control potential exposure to insect (flea and mosquito) vectors.

Summary

Viral disease in the rabbit is encountered infrequently by the clinical practitioner; however, several viral diseases were reported to occur in this species. Viral diseases that are described in the rabbit primarily may affect the integument, gastrointestinal tract or, central nervous system or may be multi-systemic in nature. Rabbit viral diseases range from oral papillomatosis, with benign clinical signs, to rabbit hemorrhagic disease and myxomatosis, which may result in significant clinical disease and mortality. The wild rabbit may serve as a reservoir for disease transmission for many of these viral agents. In general, treatment of viral disease in the rabbit is supportive in nature.

Acknowledgments

The authors would like to thank Dr. Cynthia Besch-Williford for her editorial assistance and Howard Wilson for his assistance with the figures in this article.

References

[1] Quesenberry KE, Carpenter JW. Ferrets, rabbits, and rodents clinical medicine and surgery. 2nd edition. St. Louis (MO): Saunders; 2004.
[2] Percy DH, Barthold SW. Pathology of laboratory rodents and rabbits. 2nd edition. Ames (IA): Iowa State University Press; 2001.
[3] Manning PJ, Ringler DH, Newcomer CE. The biology of the laboratory rabbit. 2nd edition. San Diego (CA): Academic Press; 1994.
[4] Harkness JE, Wagner JE. The biology and medicine of rabbits and rodents. 4th edition. Baltimore (MD): Williams and Wilkins; 1995.

[5] Karp BE, Ball NE, Scott CR, et al. Rabies in two privately owned domestic rabbits. J Am Vet Med Assoc 1999;215(12):1824–7.

[6] Weissenbock H, Hainfellner JA, Berger J, et al. Naturally occurring herpes simplex encephalitis in a domestic rabbit (*Oryctolagus cuniculus*). Vet Pathol 1997;34(1):44–7.

[7] Grest P, Albicker P, Hoelzle L, et al. Herpes simplex encephalitis in a domestic rabbit (*Oryctolagus cuniculus*). J Comp Pathol 2002;126(4):308–11.

[8] Krebs JW, Mondul AM, Rupprecht CE, et al. Rabies surveillance in the United States during 2000. J Am Vet Med Assoc 2001;219(12):1687–99.

[9] Krebs JW, Noll HR, Rupprecht CE, et al. Rabies surveillance in the United States during 2001. [erratum appears in J Am Vet Med Assoc 2003;222(4):460] J Am Vet Med Assoc 2002; 221(12):1690–701.

[10] Krebs JW, Wheeling JT, Childs JE. Rabies surveillance in the United States during 2002. J Am Vet Med Assoc 2003;223(12):1736–48.

[11] CDC. Rare rabid rabbit reported - Minnesota. Vet Public Health Notes, August, 1980. p. 1.

[12] CDC. Rabid rabbit bites a 4 year old girl - Iowa. Vet Public Health Notes, March, 1981. p. 23–4.

[13] Gallagher C. The case of the unlucky lagomorph. Vet Forum 1998;78–9.

[14] Smith JS. Rabies. Clin Microbiol Newsl, October, 1999;21:17–26.

[15] Fenner F, Wallace P. Rowe lecture. Poxviruses of laboratory animals. Lab Anim Sci 1990; 40(5):469–80.

[16] Sanarelli B. Das mycomatogene virus. Beitrag zum stadium der krankheitserreger. (The mycomatogene virus. Contribution for the stage of the pathogens) Zentralbl Bakteriol Parasitenkd Infektionsk Hyg 1898;23:865–82 (in German).

[17] Kessel JF, Prouty CC, Meyer JW. Occurrence of infectious myxomatosis in southern California. Proc Soc Exp Biol Med 1931;28:413–4.

[18] DiGiacomo RF, Mare CJ. Viral diseases. In: Manning PJ, Ringler DH, Newcomer CE, editors. The biology of the laboratory rabbit. 2nd edition. San Diego (CA): Academic Press; 1994. p. 171–204.

[19] Fenner F, Ratcliffe FN. Myxomatosis. New York: Cambridge University Press; 1965.

[20] Sobey WR. Selection for resistance to myxomatosis in domestic rabbits (*Oryctolagus cuniculus*). J Hyg (Lond) 1969;67(4):743–54.

[21] Sobey WR, Conolly D, Haycock P, et al. Myxomatosis. The effect of age upon survival of wild and domestic rabbits (*Oryctolagus cuniculus*) with a degree of genetic resistance and unselected domestic rabbits infected with myxoma virus. J Hyg (Lond) 1970;68(1): 137–49.

[22] Hurst EW. Myxoma and Shope fibroma. I. The histology of myxoma. Br J Exp Pathol 1937; 18:1–5.

[23] Chapple PJ, Bowen ETW. A note on two attenuated strains of myxoma virus isolated in Great Britain. J Hyg (Lond) 1963;61:161–8.

[24] Fenner F, Marshall ID. A comparison of the virulence for European rabbits of strains of myxoma virus recovered in the field in Australia, Europe, and America. J Hyg (Lond) 1957; 55:149–91.

[25] Patton NM, Holmes HT. Myxomatosis in domestic rabbits in Oregon. J Am Vet Med Assoc 1977;171(6):560–2.

[26] Fenner F. Classification of myxoma and fibroma viruses. Nature 1953;171:562–3.

[27] Shope RE. A transmissible tumor-like condition in rabbits. J Exp Med 1932;56:793–802.

[28] Joiner GN, Jardine JH, Gleiser CA. An epizootic of Shope fibromatosis in a commercial rabbitry. J Am Vet Med Assoc 1971;159(11):1583–7.

[29] Kilham L, Dalmat HT. Host-virus-mosquito relations of Shope fibromas in cottontail rabbits. Am J Hyg 1955;61:45–54.

[30] Kilham L, Woke PA. Laboratory transmission of fibromas (Shope) in cottontail rabbits by means of fleas and mosquitoes. Proc Soc Exp Biol Med 1953;83:296–301.

[31] Dalmat HT. Arthropod transmission of rabbit fibromatosis (shope). J Hyg (Lond) 1959;57: 1–29.

[32] McLeod C, Langlinais P. Pox virus keratitis in a rabbit. Vet Pathol 1981;18:834–6.

[33] Pulley L, Shively J. Naturally occurring infectious fibroma in the domestic rabbit. Vet Pathol 1973;10:509–19.

[34] Raflo C, Olsen R, Pakes S, et al. Characterization of a fibroma virus isolated from naturally-occurring skin tumors in domestic rabbits. Lab Anim Sci 1973;23(4):525–32.

[35] Jenkins JR. Skin disorders of the rabbit. Veterinary Clin North Am Exot Anim Pract 2001; 4(2):543–63.

[36] Shope RE, Hurst EW. Infectious papillomatosis of rabbits. J Exp Med 1933;58:607–24.

[37] Shope RE. Serial transmission of the virus of infectious papillomatosis in domestic rabbits. Proc Soc Exp Biol Med 1935;32:830–2.

[38] Hagen KW. Spontaneous papillomatosis in domestic rabbits. Bull Wildl Dis Assoc 1966;2: 108–10.

[39] Gross L. Oncogenic viruses. 3rd edition. Oxford (UK): Pergammon; 1983.

[40] Larson CL, Schillinger JE, Green RC. Transmission of rabbit papillomatosis by the rabbit tick, *Haemaphysalis leporis-palustris*. Proc Soc Exp Biol Med 1936;33:536–8.

[41] Dalmat HT. Arthropod transmission of rabbit papillomatosis. J Exp Med 1958;108:9–20.

[42] Syverton JT, Dascomb HE, Wells EB, et al. The virus-induced papilloma to carcinoma sequence. II. Carcinomas in the natural host, the cottontail rabbit. Cancer Res 1950;10: 440–4.

[43] Syverton JT. The pathogenesis of the rabbit papilloma to carcinoma sequence. Ann N Y Acad Sci 1952;54:1126–40.

[44] Weisbroth SH, Scher S. Spontaneous oral papillomatosis in rabbits. J Am Vet Med Assoc 1970;157(11):1940–4.

[45] Mews AR, Ritchie JS, Romero-Mercado CH, et al. Detection of oral papillomatosis in a British rabbit colony. Lab Anim 1972;6(2):141–5.

[46] Sundberg JP, Junge RE, El Shazly MO. Oral papillomatosis in New Zealand white rabbits. Am J Vet Res 1985;46(3):664–8.

[47] Dominguez JA, Corella EL, Auro A. Oral papillomatosis in two laboratory rabbits in Mexico. Lab Anim Sci 1981;31(1):71–3.

[48] Rdzok EJ, Shipkowitz NL, Richter WR. Rabbit oral papillomatosis: ultrastructure of experimental infection. Cancer Res 1966;26(1):160–5.

[49] Weisbroth SH. Sialocele (ranula) simulating oral papillomatosis in a domestic (*Oryctolagus*) rabbit. Lab Anim Sci 1975;25(3):321–2.

[50] Matsunaga Y, Chino F. Experimental infection of young rabbits with rabbit parvovirus. Arch Virol 1981;68(3–4):257–64.

[51] Metcalf JB, Lederman M, Stout ER, et al. Natural parvovirus infection in laboratory rabbits. Am J Vet Res 1989;50(7):1048–51.

[52] Matsunaga Y, Matsuno S, Mukoyama J. Isolation and characterization of a parvovirus of rabbits. Infect Immun 1977;18(2):495–500.

[53] Bryden AS, Thouless ME, Flewett TH. Rotavirus in rabbits. Vet Rec 1976;99:323.

[54] Conner ME, Estes MK, Graham DY. Rabbit model of rotavirus infection. J Virol 1988; 62(5):1625–33.

[55] Thouless ME, DiGiacomo RF, Neuman DS. Isolation of two lapine rotaviruses: characterization of their subgroup, serotype, and RNA electropherotypes. Arch Virol 1986;89(1–4):161–70.

[56] Tanaka TN, Conner ME, Graham DY, et al. Molecular characterization of three rabbit rotavirus strains. Arch Virol 1988;98(3–4):253–65.

[57] Castrucci G, Ferrari M, Frigeri F, et al. Isolation and characterization of cytopathic strains of rotavirus from rabbits. Arch Virol 1985;83(1–2):99–104.

[58] Petric M, Middleton PJ, Grant C, et al. Lapine rotavirus: preliminary studies on epizoology and transmission. Can J Comp Med 1978;42(1):143–7.

[59] DiGiacomo RF, Thouless ME. Age-related antibodies to rotavirus in New Zealand rabbits. J Clin Microbiol 1984;19(5):710–1.

[60] DiGiacomo RF, Thouless ME. Epidemiology of naturally occurring rotavirus infection in rabbits. Lab Anim Sci 1986;36(2):153–6.

[61] Peeters JE, Pohl P, Charlier G. Infectious agents associated with diarrhoea in commercial rabbits: a field study. Ann Rech Vet 1984;15(3):335–40.

[62] Schoeb TR, Casebolt DB, Walker VE, et al. Rotavirus-associated diarrhea in a commercial rabbitry. Lab Anim Sci 1986;36(2):149–52.

[63] Ciarlet M, Gilger MA, Barone C, et al. Rotavirus disease, but not infection and development of intestinal histopathological lesions, is age restricted in rabbits. Virology 1998;251(2): 343–60.

[64] Thouless ME, DiGiacomo RF, Deeb BJ, et al. Pathogenicity of rotavirus in rabbits. J Clin Microbiol 1988;26(5):943–7.

[65] Thouless ME, DiGiacomo RF, Deeb BJ. The effect of combined rotavirus and *Escherichia coli* infections in rabbits. Lab Anim Sci 1996;46(4):381–4.

[66] Hambraeus BA, Hambraeus LE, Wadell G. Animal model of rotavirus infection in rabbits– protection obtained without shedding of viral antigen. Arch Virol 1989;107(3–4):237–51.

[67] Castrucci G, Frigeri F, Ferrari M, et al. Comparative study of rotavirus strains of bovine and rabbit origin. Comp Immunol Microbiol Infect Dis 1984;7(3–4):171–8.

[68] Osterhaus AD, Teppema JS, van Steenis G. Coronavirus-like particles in laboratory rabbits with different syndromes in the Netherlands. Lab Anim Sci 1982;32(6):663–5.

[69] Christensen N, Fennestad KL, Bruun L. Pleural effusion disease in rabbits. Histopathological observations. Acta Pathol Microbiol Scand 1978;86(3):251–6.

[70] Alexander LK, Small JD, Edwards S, et al. An experimental model for dilated cardiomyopathy after rabbit coronavirus infection. J Infect Dis 1992;166(5):978–85.

[71] Edwards S, Small JD, Geratz JD, et al. An experimental model for myocarditis and congestive heart failure after rabbit coronavirus infection. J Infect Dis 1992;165(1):134–40.

[72] Lapierre J, Marsolais G, Pilon P, et al. Preliminary report on the observation of a coronavirus in the intestine of the laboratory rabbit. Can J Microbiol 1980;26:1204–8.

[73] Deeb BJ, DiGiacomo RF, Evermann JF, et al. Prevalence of coronavirus antibodies in rabbits. Lab Anim Sci 1993;43(5):431–3.

[74] Descoteaux JP, Lussier G. Experimental infection of young rabbits with a rabbit enteric coronavirus. Can J Vet Res 1990;54(4):473–6.

[75] Eaton P. Preliminary observations on enteritis associated with a coronavirus-like agent in rabbits. Lab Anim 1984;18(1):71–4.

[76] Ohlinger VF, Haas B, Thiel HJ. Rabbit hemorrhagic disease (RHD): characterization of the causative calicivirus. Vet Res 1993;24(2):103–16.

[77] Capucci L, Scicluna MT, Lavazza A. Diagnosis of viral haemorrhagic disease of rabbits and the European brown hare syndrome. Rev Sci Tech 1991;10(2):347–70.

[78] Campagnolo ER, Ernst MJ, Berninger ML, et al. Outbreak of rabbit hemorrhagic disease in domestic lagomorphs. J Am Vet Med Assoc 2003;223(8):1151–5.

[79] Anonymous. Rabbit calicivirus infection confirmed in an Iowa rabbitry. J Am Vet Med Assoc 2000;216:1537.

[80] Teifke JP, Reimann I, Schirrmeier H. Subacute liver necrosis after experimental infection with rabbit haemorrhagic disease virus (RHDV). J Comp Pathol 2002;126(2–3):231–4.

[81] Chasey D. Rabbit haemorrhagic disease: the new scourge of *Oryctolagus cuniculus*. Lab Anim 1997;31(1):33–44.

[82] Wirblich C, Meyers G, Ohlinger VF, et al. European brown hare syndrome virus: relationship to rabbit hemorrhagic disease virus and other caliciviruses. J Virol 1994;68(8):5164–73.

[83] Gavier-Widen D. Morphologic and immunohistochemical characterization of the hepatic lesions associated with European brown hare syndrome. Vet Pathol 1994;31(3):327–34.

[84] Nauwynck H, Callebaut P, Peeters J, et al. Susceptibility of hares and rabbits to a Belgian isolate of European brown hare syndrome virus. J Wildl Dis 1993;29(2):203–8.

[85] McColl KA, Merchant JC, Hardy J, et al. Evidence for insect transmission of rabbit haemorrhagic disease virus. Epidemiol Infect 2002;129(3):655–63.

[86] Dressler RL, Ganaway JR, Storm GL, et al. Serum antibody prevalence for *Herpesvirus sylvilagus, Bacillus piliformis* and California serogroup arboviruses in cottontail rabbits from Pennsylvania. J Wildl Dis 1988;24(2):352–5.

[87] Schmidt SP, Bates GN, Lewandoski PJ. Probable herpesvirus infection in an eastern cottontail (*Sylvilagus floridanus*). J Wildl Dis 1992;28(4):618–22.

[88] Spieker JO, Yuill TM. *Herpesvirus sylvilagus* in cottontail rabbits: evidence of shedding but not transplacental transmission. J Wildl Dis 1977;13(1):85–9.

[89] Hesselton RM, Yang WC, Medveczky P, et al. Pathogenesis of *Herpesvirus sylvilagus* infection in cottontail rabbits. Am J Pathol 1988;133(3):639–47.

[90] Yang WC, Hesselton RM, Sullivan JL. Immune responses to *Herpesvirus sylvilagus* infection in cottontail rabbits. J Immunol 1990;145(6):1929–33.

[91] Kerr PJ, Best SM. Myxoma virus in rabbits. Rev Sci Tech 1998;17(1):256–68.

[92] Calvete C, Estrada R, Villafuerte R, et al. Epidemiology of viral haemorrhagic disease and myxomatosis in a free-living population of wild rabbits. Vet Rec 2002;150(25):776–82.

[93] Marlier D, Mainil J, Sulon J, et al. Study of the virulence of five strains of amyxomatous myxoma virus in crossbred New Zealand White/Californian conventional rabbits, with evidence of long-term testicular infection in recovered animals. J Comp Pathol 2000; 122(2–3):101–13.

[94] Fountain S, Holland MK, Hinds LA, et al. Interstitial orchitis with impaired steroidogenesis and spermatogenesis in the testes of rabbits infected with an attenuated strain of myxoma virus. J Reprod Fertil 1997;110(1):161–9.

[95] Licon Luna RM. First report of myxomatosis in Mexico. J Wildl Dis 2000;36(3):580–3.

[96] Torres JM, Sanchez C, Ramirez MA, et al. First field trial of a transmissible recombinant vaccine against myxomatosis and rabbit hemorrhagic disease. Vaccine 2001;19(31):4536–43.

[97] Psikal I, Smid B, Rodak L, et al. Atypical myxomatosis–virus isolation, experimental infection of rabbits and restriction endonuclease analysis of the isolate. J Vet Med B Infect Dis Vet Public Health 2003;50(6):259–64.

ELSEVIER
SAUNDERS

Vet Clin Exot Anim 8 (2005) 139–160

VETERINARY
CLINICS
Exotic Animal Practice

Viral diseases of ferrets

Isabelle Langlois, DMV, Dipl. ABVP (Avian practice)

*Médecine Zoologique, Centre Hospitalier Universitaire Vétérinaire, Faculté de Médecine
Vétérinaire, Université de Montréal, C.P. 5000, St-Hyacinthe, Québec J2S 7C6, Canada*

With the continually growing popularity of ferrets, practitioners will encounter individuals that suffer from viral disease more frequently. Ferrets have been used extensively as animal models to study a variety of viral diseases, many of which only occur following experimental infection. This article provides a review of viral diseases that occur naturally in ferrets (Table 1). The etiology, transmission, and pathogenesis of each virus is discussed. There is special emphasis on clinical signs, diagnostic tests, and methods to control and prevent these conditions.

Distemper

Distemper is caused by an RNA virus of the genus Morbillivirus in the Paramyxoviridae family. There is one serotype of canine distemper virus (CDV) with several strains that produce different clinical presentations [1]. Distemper probably is the most serious infectious disease of ferrets; mortality rates approach 100%. Infection of ferrets with CDV is uncommon because of the availability of effective vaccines and client education on the importance of vaccination. Unvaccinated dogs and wild Canidae, Mustelidae and Procyonidae may serve as reservoirs of the disease [1]. A CDV epizootic was reported in a black-footed ferret (*Mustela nigripes*) colony in the fall of 1985 in Wyoming. The source of infection was not identified, but several badgers and coyotes in the area had CDV-neutralizing antibodies [2].

Pathogenesis

CDV is transmitted by aerosol exposure to infected body fluids and by direct contact with infected animals and contaminated fomites. The virus is shed in ocular and nasal secretions, saliva, urine, and feces [1]. Viral

---

E-mail address: isabelle.langlois@umontreal.ca

Table 1
List of naturally occurring viruses of domestic ferrets

Viruses	Family	Nucleic acid	Envelope	Size (nm)
Distemper virus	Paramyxoviridae	ssRNA	yes	150–300
Coronavirus	Coronaviridae	ssRNA	yes	60–220
Influenza virus	Orthomyxoviridae	ssRNA	yes	90–120
Rabies virus	Rhabdoviridae	ssRNA	yes	70–85 × 130–380
Rotavirus	Reoviridae	dsRNA	no	80–130
Parvovirus	Parvoviridae	ssDNA	no	18–26
Infectious bovine rhinotracheitis virus	Herpesviridae	dsDNA	yes	46–48

Abbreviations: ds, double stranded; ss, single stranded.

shedding begins approximately 7 days postinfection [3]. The primary site of viral replication is the respiratory epithelium and lymphoid tissue of the nasopharynx [1,4]. The virus disseminates by way of peripheral blood leukocytes to the liver, kidneys, gastrointestinal tract, urinary bladder, and brain. Viremia has been documented 2 days postinoculation or infection and persists until the virus is neutralized by antibodies or the animal dies [5]. Gastric hypochlorhydria has been associated with CDV infection in ferrets [6]. The lack of gastric acid was hypothesized to be due to direct action of the virus on gastric mucosa or was secondary to the viral effect on the central nervous system.

Clinical signs

Infected ferrets become symptomatic after an incubation period of 7 to 10 days [1,7,8]. In ferrets, natural CDV infection most often consists of a catarrhal phase followed by a fatal neurotropic phase [9]. The initial phase is characterized by anorexia, pyrexia, conjunctivitis, and serous nasal discharge. An erythematous, pruritic rash appears on the chin (Fig. 1) and eventually spreads to the inguinal area [1,9]. Hyperkeratosis of the footpads (Fig. 2) occurs inconsistently in ferrets [1]. Melena may be observed early in the course of the disease [1,9]. Some ferrets die during the catarrhal phase secondary to bacterial infections, such as pneumonia. Clinical signs that are seen in the neurotropic phase include hyperexcitability, muscle tremors, hypersalivation, seizures, and coma. Ferrets die 12 to 16 days after being infected with ferret-adapted CDV [1,8]. The disease has a longer course when ferrets are infected with a wild canine strain; death occurs 21 to 35 days postinfection [1]. Sometimes a decrease in temperature is observed by the time the animals become moribund [4].

Diagnosis

A tentative diagnosis of canine distemper is based on the presence of typical clinical signs, severe leukopenia, a history of potential exposure to the virus, and questionable vaccination. A study demonstrated that virulent

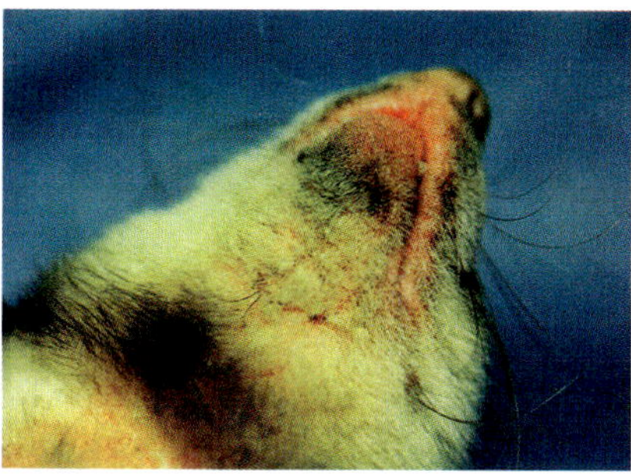

Fig. 1. An erythematous, cutaneous rash on the chin of a ferret with canine distemper virus infection. (*From* Blair EM, Chambers MA, King HA. Treating distemper in a young ferret. Vet Med 1998;98(7):656; with permission.)

and attenuated CDV strains caused a severe leukopenia 1 week after infection [4]. Leukocyte counts from animals that were infected with virulent strains remained low, whereas those of animals that received attenuated strains progressively increased to reach preinfection, or near preinfection, values 35 days postinfection.

The diagnosis of canine distemper is confirmed routinely by an immunofluorescence test on peripheral blood smear, buffy coat, or conjunctival scrapings in live animals. In experimentally-infected ferrets, a reverse transcriptase–polymerase chain reaction (RT-PCR) has been used to detect the virus in peripheral blood [10]. Recently, nested-PCR was found to be a sensitive and specific method for diagnosis of CDV infection [11].

On postmortem examination, the gross lesions correspond to the aforementioned clinical signs. On hematoxylin and eosin–stained sections, round eosinophilic intracytoplasmic and occasional intranuclear inclusion bodies are present in epithelial cells of the trachea, bronchi, and urinary tract [1]. Other tissues, such as skin, gastrointestinal tract, salivary and adrenal glands, spleen, lymph nodes and brain, have been reported to contain inclusions [1].

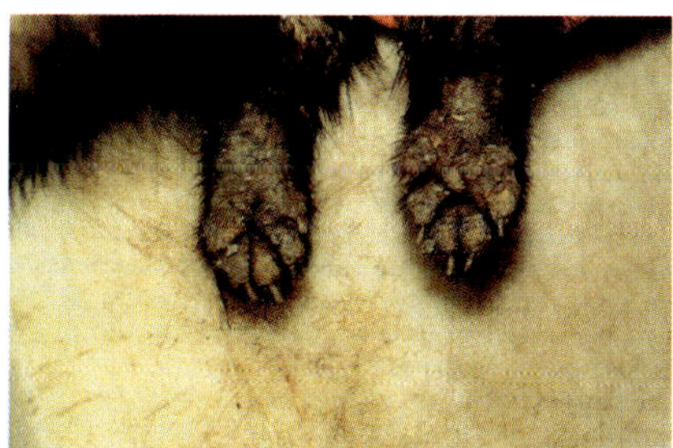

Fig. 2. Hyperkeratosis on the footpads of a ferret with canine distemper virus infection. (*From* Blair EM, Chambers MA, King HA. Treating distemper in a young ferret. Vet Med 1998; 98(7):657; with permission.)

Treatment and prevention

Ferrets that have signs that are suggestive of canine distemper infection should be placed in isolation. Supportive care, including fluid therapy, systemic and ophthalmic antibiotics, gavage feeding, and bathing with antipruritic shampoo, may be initiated. Administration of immune anti-CDV serum also may be considered. Thorough daily disinfection of the environment is indicated. In all cases, prognosis is grave. The virus is inactivated by heat, visible light, and commonly used disinfectants (0.75% phenol, 0.2% roccal, 2%–5% sodium hydroxide, and 0.1% formalin) [1,3]. Some investigators reported that out of more than 1000 cases of distemper, not a single ferret survived the infection [8]. Blair et al [12] treated distemper in a 7-month-old ferret while in the catarrhal phase. The ferret's life span and quality of life were improved with treatments, but the neurotropic phase could not be prevented and the animal was euthanized. When faced with an outbreak of canine distemper in a ferret colony, euthanasia and repopulation following disinfection is recommended.

Distemper is prevented best by vaccination. Two vaccines are approved for use in ferrets in North America—Fervac-D Canine Distemper Vaccine (United Vaccines Inc., Madison, Wisconsin) and Purevax Ferret Distemper Vaccine (Merial, Athens, Georgia and Montréal, Quebec, Canada). Fervac-D is a modified-live virus vaccine of chick cell origin. A 5.9% incidence of anaphylactic reaction has been reported with this vaccine [13]. Adverse events developed within 25 minutes and were characterized by hyperemia, hypersalivation, and vomiting. Purevax is a lyophilized vaccine of a recombinant canary pox vector that expresses the HA and F glycoproteins of CDV. The manufacturer uses the term "ferret distemper" vaccine on the product label to prevent confusion with their canine products; however, there is no such thing as a "ferret distemper virus" and ferrets are infected by CDV. The absence of adjuvant or entire distemper virus decreased postvaccination risks and the manufacturer reports an incidence of 0.3% reversible anaphylactic reactions in its field safety trials. Another modified-live canine distemper vaccine attenuated in primate cell line (Galaxy D, Schering Plough Animal Health, Omaha, Nebraska) has been studied in ferrets. This vaccine was effective in preventing canine distemper in young ferrets that were challenged with virulent CDV after two vaccine inoculations [14]; however, the duration of immunity and the incidence of vaccine reactions are unknown. Clinical use of this vaccine is extra label (not approved for use in ferrets by USDA) and requires informed owner consent. Postvaccinal distemper infections were reported in black-footed ferrets that were vaccinated with chicken embryo–tissue culture–origin CDV vaccine and in domestic ferrets that were vaccinated with canine cell products [15,16]. Therefore, vaccination of ferrets with these products must not be performed.

Most ferrets that are obtained from pet stores at a young age received only one dose of CDV vaccine before leaving the breeding facility. Repeated inoculations are required because maternal antibodies may interfere with proper immune response to CDV vaccine antigens. Ferret kits should be vaccinated at 8 weeks of age and every 3 weeks for a total of 3 vaccinations [17,18]. Yearly booster vaccines are recommended [17]. Observation of the ferret for 25 minutes following vaccination is advisable. In the event that a vaccine reaction occurs, administration of fluids, oxygen, antihistamine, and epinephrine may be indicated.

Parvovirus

Parvovirus strains of varying virulence and immunogenicity cause Aleutian disease. The disease was first reported in mink in the 1940s and got its name because mink that were homozygous for the Aleutian (blue) gene were affected most severely [1]. The infection was documented in ferrets in the late 1960s [19]. At least three separate strains of Aleutian disease virus (ADV)—distinct from the mink strains—have been identified in ferrets [20,21]. The ferret strains are believed to be mutant strains of the mink parvovirus; the hypervariable capsid region of the ferret strains of ADV is similar to that of the mink parvovirus [22]. Ferrets can be infected with the mink virus [20,21] and ferret ADV can infect mink; however, the virulence is lower compared with mink that are infected with mink strains [21,23,24].

Pathogenesis

Transmission of the virus may occur by aerosolization; by direct contact with urine, feces, saliva, and blood; or by contact with fomites [1,24]. Vertical transmission of ADV occurs in mink [1,25]. The lesions that are caused by ADV infection are immune mediated, but the mechanism by which ADV interferes with the immune system is unknown. The severity of disease depends on the origin (mink or ferret) of the ADV strain that is involved as well as the immune status and genotype of the infected individual [25].

Minks that are infected with mink ADV strains deposit immune complexes in various organs that result in glomerulonephritis, bile duct proliferation, and arteritis [26]. Mink kits from antibody-free jills die from acute interstitial pneumonia when infected with virulent ADV [27]. Affected individuals are immunosuppressed, and therefore, are more susceptible to influenza, viral enteritis, and distemper [1,26]. Ferrets that are inoculated with ferret-adapted ADV exhibit marked, persistent hyperglobulinemia and periportal lymphocytic infiltrates of the liver [21]; however, immunocompetent adult ferrets that are infected experimentally can develop a persistent infection without clinical disease [21]. Mink strains cause milder lesions and only a moderate elevation in gamma globulins in ferrets [1].

Clinical signs

Most ferrets that show clinical signs are between 2 and 4 years old. Ferrets can be infected for years before clinical symptoms are noted [26].

Any situation that leads to immunosuppression may play a role in the development of clinical disease. The clinical presentation of infected ferrets varies. Some ferrets that are infected with ADV die without clinical signs in good body condition [24]. Generally, ferrets show signs of a chronic wasting disease with progressive weight loss, malaise, and melena [1]. Acute dyspnea was described in one report [28]. Central nervous system signs, such as tremors, ataxia, paralysis, and convulsions, also have been reported [25,29–32]. Affected animal also may present with fecal and urinary incontinence [25,29].

Diagnosis

A presumptive diagnosis can be made based on a high gamma globulin concentration, the history, and clinical signs. Although hypergammaglobulinemia is considered to be pathognomonic in mink, this feature is not always present in infected ferrets [28,29]. Serum protein electrophoresis often shows that gamma globulins account for more than 20% of the total protein concentration [1,21].

Other clinical pathologic findings of infected ferrets are variable. Anemia, possibly attributable to hemolysis and decreased erythrocyte production, may be present [25]. Biochemical abnormalities, such as azotemia and elevated liver enzymes, may be seen according to damage that results from immune complex deposition. Proteinuria and urinary casts that are secondary to kidney damage also may be observed [28].

Diagnosis of Aleutian disease is confirmed antemortem with a positive serum titer coupled with hypergammaglobulinemia or lymphoplasmacytic inflammation in tissue biopsy samples. Two serologic tests are available for ADV testing—counterimmunoelectrophoresis (CEP or CIEP) (United Vaccines, Madison, Wisconsin) and ELISA (Avecon Diagnostics, Bath, Pennsylvania). The specificity and sensitivity of the ELISA test has not been investigated in ferrets. Therefore, results must be interpreted with caution. The CEP test is used routinely in ferrets and is an effective method for identifying ferrets that have ADV antibodies [25,30–33]. Presence of antibodies without clinical disease for extended periods is possible [34]. Ferrets probably develop persistent and nonpersistent, nonprogressive forms of ADV infection similar to mink [25,35].

Detection of ADV DNA by in situ hybridization has been performed, but this method is not practical to screen for this condition [36]. Recently, polymerase chain reaction amplification of part of the capsid gene that is specific to ADV and restriction fragment length polymorphism to distinguish the ferret types of ADV from the mink types of ADV are valuable, time-saving assets for diagnosis of this infection in ferrets [20].

At necropsy, infected ferrets may have few or no gross lesions. Hepatosplenomegaly, splenomegaly, and mesenteric lymphadenopathy have been reported [23,24]. The most consistent histologic findings of

ADV infection in ferrets are periportal infiltration of the liver by plasma cells, lymphocytes, and macrophages with stimulated lymphoid tissues [23,24]. Bile duct hyperplasia, periportal fibrosis, and membranous glomerulonephrosis have been documented [23,24,28]. In individuals that present with neurologic signs, perivascular lymphocytic cuffing in the brain and spinal cord (Fig. 3) and lymphoplasmacytic meningitis may be observed [25,30,31].

Treatment and control

There is no definitive treatment of Aleutian disease in ferrets. Symptomatic ferrets may benefit from use of anti-inflammatory medication or immunosuppressive drugs, such as prednisone and cyclophosphamide. In mink, treatment with cyclophosphamide has been used to control infections for up to 16 weeks, but virus titers did not decrease [37]. Administration of gamma globulin–containing ADV antibody may be considered because it contributed to decreased mortality rates in mink kits [38].

Control of Aleutian disease is dependent on testing, cessation of breeding, and isolation of seropositive ferrets. All seropositive ferrets should be considered to be potential sources of ADV and should be isolated from seronegative ferrets [1]. In mink farms, testing by CIEP and removing all individuals that have ADV antibodies has been an efficient method to eradicate the disease [39]. This approach, coupled with thorough disinfection, should be considered in any facility with a large number of ferrets. Formalin, sodium hydroxide, and a phenolic disinfectant were efficacious against ADV in the presence of organic material [40].

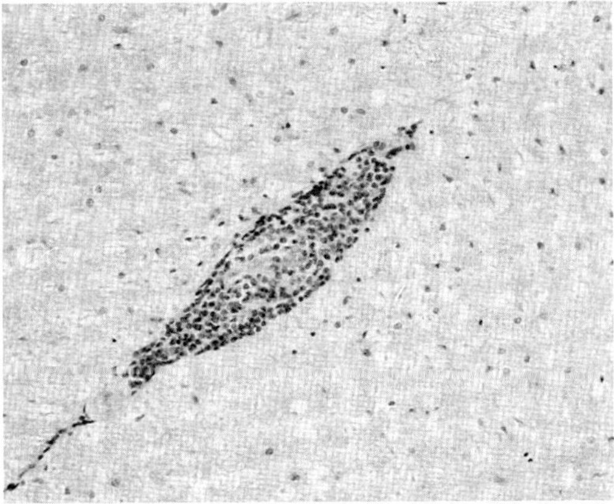

Fig. 3. Severe perivascular cuffing with mononuclear cells in the spinal cord of a ferret with Aleutian disease. (*From* Welchman D de B, Oxenham M, Done SH. Aleutian disease in domestic ferrets: diagnostic findings and survey results. Vet Rec 1993;132:481; with permission.)

There is no vaccine to prevent Aleutian disease and vaccination probably is contraindicated because of the immune-mediated nature of this condition. In mink, vaccination exacerbated the severity of Aleutian disease [41].

Coronavirus

In the late 1980s, a novel diarrheal disease that affected domestic ferrets was reported by pet owners and ferret breeders in the mid-Atlantic area of the United States. Since that time, this condition has been diagnosed throughout North America and in several other countries. The disease was named epizootic catarrhal enteritis (ECE) on the basis of similarities to the epizootic catarrhal gastroenteritis of mink [42]. ECE of mink is caused by a coronavirus that is related to transmissible gastroenteritis virus of pigs [43]. Research strongly implicates a coronavirus as the causative agent of ECE because: (1) microscopic lesions that were consistent with intestinal coronavirus infection were detected consistently in diseased ferrets; (2) coronavirus particles were identified in the feces and enterocytes, but no other viruses could be identified; and (3) immunohistochemical staining of jejunum showed coronavirus antigens in affected ferrets, but not in healthy individuals [44].

Clinical presentation

The disease is characterized by high transmissibility, high morbidity, and a low mortality rate. Ferrets show signs of lethargy and anorexia within 48 to 72 hours postinfection. Vomiting is the first sign of gastrointestinal disease in most ferrets, but it may go unnoticed by some owners because it subsides within hours [44]. Subsequently, profuse green, bile-tinged diarrhea with a variable amount of mucus develops (Fig. 4). The stool's appearance is responsible for the term "green slime disease" that also is used to describe this condition [45].

The severity of clinical signs is highly variable; older ferrets that have concurrent diseases, such as insulinoma, long-standing infection with *Helicobacter mustelae*, or adrenal neoplasia, are prone to develop severe clinical signs. Ulcerations of the intestinal wall may occur which leads to the presence of blood in the feces. Young ferrets tend to have mild or subclinical infection. The hypersecretory phase of uncomplicated ECE often resolves within 5 to 7 days in healthy young animals. In some ferrets, this phase may be followed by a period of maldigestion or malabsorption of variable duration secondary to persistent lymphocytic inflammation of the intestinal wall. The feces become yellowish in color and contain grainy material that resembles bird seed.

Diagnosis

ECE often can be diagnosed solely on the basis of characteristic historical findings and clinical signs [44]. Thorough collection of the history data often

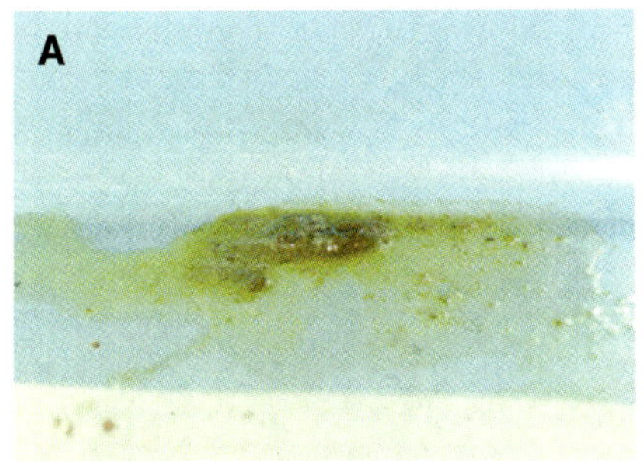

Fig. 4. (*A, B*) A. Profuse green bile-tinged diarrhea from the young ferret with epizootic catarrhal enteritis shown in Fig. B. (Courtesy of C. Greenacre, DVM, Knoxville, Tennessee.)

reveals exposure to an asymptomatic ferret 48 to 72 hours before the onset of clinical signs. Generally, clinicopathologic findings are nonspecific. Inanition may cause increased serum activity of alanine aminotransferase and alkaline phosphatase secondary to mobilization of peripheral fat stores to the liver, with resultant hepatocellular swelling [1,44,45]. Hypoalbuminemia may develop as a result of enteritis and malabsorption in chronically affected individuals. Leukocytosis may be present in ferrets that have concurrent bacterial infection or gastric ulcers. Definitive diagnosis of coronavirus infection often is difficult. Coronavirus-like particles may be identified in the feces by electron microscopy during the acute phase of the disease process. Characteristic histologic lesions that are seen in intestinal coronavirus infection, such as lymphocytic enteritis with villus atrophy, fusion, blunting and vacuolar degeneration or necrosis of the apical epithelium, may be identified on intestinal biopsy or necropsy specimens [44].

Treatment

Ferrets may become dehydrated rapidly during the hypersecretory phase of infection. Dehydration and electrolyte imbalances need to be addressed. Fluids that are supplemented with dextrose and electrolytes may be administered orally, subcutaneously, or intravenously, according to the degree of dehydration. Antibiotic may be indicated to prevent secondary bacterial infection, particularly in cases with suspected intestinal ulcerations. In these cases, administration of sucralfate and an H_2 antagonist (eg, cimetidine) also are beneficial. Sucralfate requires an acidic environment to be effective, so it should be given at least 30 minutes before an H_2 antagonist. Syringe feeding with a highly digestible diet (Science Diet A/D, Hill's Prescription Diet) may be indicated if anorexia persists. If clinical signs that are suggestive of maldigestion develop, oral administration of prednisone may be indicated. Use of an anti-inflammatory dose of prednisone for 1 to 4 weeks, was followed by gradual tapering of the dose successfully.

Influenza virus

Influenza viruses belong to the class Orthomyxoviridae. Human influenza types A and B are pathogenic to ferrets [1]. Ferrets also are susceptible to avian, seal, equine, and swine influenza A viruses, although only human, avian, and swine strains induce clinical signs [46–50]. Infection with influenza B virus less frequently results in illness and is associated with a milder clinical course [1]. Transmission of influenza virus from human to ferrets and from ferrets to humans was documented in the 1930s [51,52]. Ferrets are used extensively as an animal model for influenza virus pathogenesis and immunity studies because their biologic response to influenza infection is similar to that of humans [53,54].

Pathogenesis and clinical signs

Influenza virus is transmitted by aerosol droplets from an infected individual, either to a human or a ferret. After intranasal inoculation, the virus localizes and replicates in great numbers within the nasal mucosa [55]. Following a short incubation period, the body temperature increases and then decreases approximately 48 hours later [56]. Transmission of the virus begins at the height of pyrexia and continues for 3 to 4 days [53]. As in humans, the disease is characterized by upper respiratory signs. Clinical signs appear 48 hours postinfection and include anorexia, malaise, fever, sneezing, and serous nasal discharge [1,56]. Infection usually is mild in adult ferrets compared with neonates who can be severely ill [57]. Conjunctivitis, photosensitivity, and unilateral otitis also may be seen [1,58].

Influenza infection may involve the lower respiratory tract in some susceptible animals [1,59]. Usually, influenza virus is confined to the bronchial and bronchiolar tissues [53,60]. The disease may be fatal in 1- to 2-day-old ferret kits secondary to bronchiolitis, pneumonia, and aspiration of material from the upper respiratory tract [57,61,62]. Lancefield group C hemolytic streptococci have been involved in secondary bacterial pneumonia [56].

Influenza virus may infect the intestinal epithelium and cause limited enteritis [50,63]. Hepatic dysfunction also has been reported in ferrets that were infected experimentally with influenza [64]. Neurologic symptoms, including ataxia, hind-limb paresis, and torticollis, were reported in ferrets that were infected experimentally with avian influenza A (H5N1) viruses that were isolated from the 1997 outbreaks of disease in domestic poultry markets in Hong Kong [50,65].

Diagnosis

Generally, the diagnosis of influenza infection is based on the presence of compatible clinical signs, a history of exposure to infected individuals, and recovery from illness within 7 to 10 days. The differential diagnosis of any ferrets that has upper respiratory signs should include canine distemper. The usually mild and brief nature of influenza infections help to distinguish it from distemper. The use of virus isolation or hemagglutinin-inhibiting antibody titers on acute and convalescent serum samples rarely is needed for a diagnosis [1]. An enzyme-linked immunosorbent assay has been used to detect antibodies against influenza A and may be used to obtain a diagnosis rapidly [66].

Clinical pathologic findings may present abnormalities. Studies demonstrated an elevation in the neutrophil:lymphocyte ratio in the peripheral blood [50,67]. Transient lymphopenia, with a loss of 60% to 65% of peripheral blood lymphocytes 3 days postinfection was reported experimentally with avian influenza A (H5N1) [50]. Plasma biochemical values generally are within reference ranges, but increases in concentrations of creatinine, blood urea nitrogen, potassium, albumin, and alanine aminotransferase were reported in some infected ferrets [64].

Treatment

In most cases, infected ferrets can be treated at home. Owners should be instructed to let their ferret rest until fully recovered. Offering affected animals their favorite diet, highly palatable food (chicken baby food, beef baby food), or a highly energetic diet (Science Diet A/D, Hill's Prescription Diet) is indicated. Force feeding and offering water by syringe can be performed as needed. Treatments to relieve clinical symptoms should be done on a case by case basis. If coughing is persistent, a pediatric cough suppressant without alcohol (at the pediatric dosage on a per weight basis) has been used [59]. To alleviate nasal congestion, the use of an antihistamine, such as diphenhydramine (2–4 mg/kg, by mouth, every 8 to 12 hours) [8,59], or intranasal delivery of phenylephrine may help [68]. Antibiotics may be indicated to control secondary bacterial infections of the respiratory tract. Neonates typically succumb to secondary bacterial infections. Therefore, antibiotics may be useful in these patients to reduce mortality [69].

The use of non steroidal anti-inflammatory drugs to control fever is of questionable benefit because fever seems to be important in restricting the severity of infection [70]. Experimentally, ferrets who received aspirin had cooler body temperature, but they shed more virus and their viral levels decreased less rapidly compared with ferrets that were not treated with an antipyretic.

Administration of antiviral medication has been studied in ferrets. Amantadine hydrochloride (6 mg/kg, by mouth, every 12 hours) (Symmetrel, Bristol-Myers Squibb Canada, Montreal, Quebec, Canada) has been effective in treating ferrets that have influenza [71]. Zanamivir (12.5 mg/kg) (Relenza, Glaxo Wellcome, Mississauga, Ontario, Canada), given as a one-time intranasal dose was able to prevent influenza infection [72]. Administration of amantadine in ferrets rapidly produces antiviral resistance, but use of zanamivir does not [73].

Prevention and control

Vaccination of ferrets against influenza virus generally is not recommended because it is usually a mild disease and the antigenic variation of the virus complicates vaccination [1,59]. Experimentally, ferrets who recovered from influenza infection remained resistant to infection with the same strain for 5 weeks following initial infection [74]. Ferret kits are protected from disease by milk-derived antibodies in immunized females [75].

Controlling influenza infection resides in avoiding exposure of susceptible ferrets to infected individuals, either ferret or human. Owners should be advised to minimize contact with their ferrets if they have a respiratory infection and should be sure to wash their hands thoroughly before changing the animal's cage, food, and water. Veterinarians who have respiratory infections may consider wearing a mask and gloves.

Rhabdovirus

Rhabdovirus causes rabies, an acute and almost invariably fatal disease that affects many mammals and humans. Over the past several years, there have been numerous reports of ferret bite injuries, including unprovoked attacks on infants and small children [76–78]. These reports brought great controversy over the acceptability of keeping ferrets as pets with regard to the potential risk for rabies. This was because the period of viral shedding in the animal's saliva—before the onset of recognizable signs—was unknown at that time [1,79,80]. To the author's knowledge, there is no reported case of human rabies secondary to a ferret bite. Since 1958, less than 30 cases of rabies in domestic ferrets have been reported to the U.S. Centers for Disease Control [81,82]. One of theses cases was attributed to vaccinating a ferret with modified-live virus rabies vaccine [1].

Pathogenesis

Rabies virus must contact nerve endings and enter nerve fibers before infection occurs. Exposure to rabies virus does not always lead to productive infection [81–83]. Host response to rabies virus is influenced by the rabies virus variant, the viral dose, the route of transmission, the host species, and individual variations [84]. Infection occurs primarily by contact of nerve endings with infected saliva from a rabid animal as a result of a bite wound. Contact with the conjunctiva or olfactory mucosa also can result in transmission [84]. Transmission of rabies through ingestion of an infected mouse was unsuccessful [85].

The pathogenesis of rabies in ferrets has been studied using European red fox rabies variant, North Central skunk rabies variant, and raccoon rabies variant [81–83]. The mean incubation period is approximately 1 month [81,83]. Clinical signs reported include ascending paralysis, ataxia, tremors, paresthesia, hyperactivity, anorexia, cachexia, bladder atony, constipation, fever, and hypothermia [81–83]. Two of 19 rabid ferrets that were inoculated with raccoon rabies variant showed aggressive behavior [83]. Signs were reported to be mild in ferrets that were inoculated with the European red fox rabies variant [82]. Mean morbidity period was 4 to 5 days [81,83]. Mortality in ferrets that were inoculated with the European red fox variant was dose dependent [82]. In the study that used skunk variant, ferret susceptibility was dose dependent and the incubation period was inversely proportional to dose [81]. In contrast, ferrets that were inoculated with raccoon variant were only moderately susceptible—regardless of dose—and there was no correlation found between viral dose and incubation period [83]. Ferrets who survived experimental infection remained clinically normal except for one ferret that was given skunk rabies variant and had severe paralytic sequelae [81–83].

The immunologic response of ferrets to rabies virus infection seems to vary depending on the rabies virus variant that was inoculated. Virus neutralizing antibodies were demonstrated in 14 of 33 (42.4%) rabid ferrets that were inoculated with skunk rabies variant compared with only 2 of 19 (10.5%) rabid ferrets that were given raccoon rabies variant [81,83]. Also, among ferrets that survived the infection, the proportion of seropositive cases was greater in ferrets that were given skunk rabies variant [81,83].

Ferrets may or may not excrete rabies virus in their saliva depending on the virus variant to which they have been exposed. Viral shedding was not documented in the saliva 120 days postinoculation with European red fox variant [82]. Rabies virus was not detected in the saliva of any ferrets that were given skunk rabies variant, but it was isolated from the submaxillary salivary gland of one rabid ferret that was euthanized [81]. Shedding of rabies virus in the saliva was documented in ferrets that were inoculated with the raccoon rabies variant [83]. Rabies virus was isolated from the salivary glands of 63% of rabid ferrets and 47% shed rabies virus in their saliva. Initial viral excretion ranged from 2 days before the onset of clinical signs to 6 days after the onset. Viral shedding also was documented in 1 of 23 ferrets that were inoculated with a rodent strain of rabies virus [86].

Diagnosis

Rabies should be included in the differential diagnoses of any ferret that has an unexpected onset of paralysis or acute personality change, especially if the ferret is unvaccinated, has access to outdoors, or lives in an area that is experiencing an epizootic of rabies. A ferret that is suspected of having rabies should be euthanized and its head should be submitted to the appropriate laboratory. Generally, the diagnosis is based on direct immunofluorescent antibody testing of brain tissue [1,81,83]. Confirmation may be established by intracerebral inoculation of suckling mice or inoculation of tissue culture with homogenized brain tissue [1].

Prevention and control

The protective efficacy of killed rabies vaccine was demonstrated in ferrets that were vaccinated once subcutaneously with an inactivated rabies virus and were challenged 1 year later with street virus of fox origin. Vaccinated ferrets had a survival rate of 89% compared with less than 6% for unvaccinated controls [87]. Domestic ferrets that were vaccinated with a commercially inactivated rabies vaccine showed rapid induction of virus-neutralizing antibody—a seroconversion that persisted at least 7 months [88]. Based on these studies, killed rabies vaccines were approved for immunizing ferrets [89,90] (Table 2). Ferrets should be vaccinated once at 3 months of age or older and then annually thereafter [87]. An animal is considered to be immunized if the primary vaccination was administered at least 28 days previously [89]. Because a rapid anamnestic response is expected, an animal is considered to be vaccinated currently immediately after booster vaccination [89].

Table 2
Killed rabies vaccines licensed for use in ferrets in the United States and Canada

Trade name	Country where licensed	Manufacturer	Dosage	Route	Recommended schedule
Imrab 3 TF	Canada United States	Merial	1 mL	Subcutaneous	3 months, then annually
Imrab 3	Canada United States	Merial	1 mL	Subcutaneous	3 months, then annually
Prorab	Canada	Intervet Canada	1 mL	Subcutaneous	3 months, then annually

Data from National Association of State Public Health Veterinarians. Compendium of animal rabies prevention and control, 2003. Morb Mortal Wkly Rep 2003;52(RR-5):14; Rabies Vaccines licensed in Canada. Canadian Food Inspection Agency, Animal Health and Production Division, Veterinary Biologics Section. Available at: http://www.inspection.gc.ca/english/anima/vetbio/prod/rabrage.shtml. Accessed October 2004.

Local injection site reactions were reported to develop frequently with rabies vaccine [91]. Studies in cats showed that rabies vaccines more consistently produce granulomatous inflammation at vaccine sites, although leukemia virus vaccines are incriminated more often in the pathogenesis of vaccine-associated sarcomas [92–94]. There is only one report of vaccine-associated sarcoma in a ferret [91]. The ferret had been vaccinated for distemper and rabies on an annual basis in the dorsal area of the neck or interscapular area, so it is impossible to determine from which vaccine the tumor arose. Practitioners should consider establishing a protocol in regard to the site of administration of rabies and distemper vaccines. This would allow the determination of which vaccine may be involved in the event that a tumor develops.

A study demonstrated that the cellular response to the canary pox–vectored rabies vaccine in ferrets was much milder than to the adjuvanted rabies vaccines [95]. Consequently, its future use may be associated with a decreasing number of local vaccine reactions. Also, because persistence of lymphocytes and macrophages has been suggested to play a role in the pathogenesis of vaccine-associated sarcomas, the canary pox–vectored vaccine would be less likely to be involved in the oncogenesis of these tumors. Although it shows great promise, practitioners should remember that canary pox–vectored rabies vaccine is not approved for use in ferrets in North America at this time.

To the author's knowledge, there is only one report of anaphylactic reaction in a ferret following administration of an inactivated rabies vaccine [13,96]. This ferret previously had an anaphylactic reaction after receiving a distemper and rabies vaccine simultaneously. Observation of ferrets for 25 minutes following rabies vaccination is advisable.

Unvaccinated ferrets that are exposed to a rabid animal should be euthanized immediately [89,97]. If the owner refuses, the animal should be quarantined strictly for 6 months and vaccinated 1 month before being released [89,97]. Ferrets that are vaccinated currently should be revaccinated immediately [89,97].

If a healthy ferret bites a person, it should be confined and observed for 10 days [89,97]. The animal needs to be evaluated by a veterinarian at the first sign of illness during confinement. If signs that are suggestive of rabies develop, the animal should be euthanized and tested for rabies [89,97].

Rotavirus

Rotavirus belongs to the family Reoviridae. All rotaviruses are unique because they possess double-stranded RNA genome. Generally, these viruses cause diarrhea in young animals and children, but they also can occur in older individuals [98]. An atypical rotavirus was isolated from neonatal ferrets (*Mustela putorius furo*) that had diarrhea at a large commercial farm in the United States. The disease was recognized at the ferret farm for several years and was referred as "ferret kit disease" [1]. Partial characterization identified this virus as an atypical rotavirus, based on the lack of rotavirus group A common antigen and on its distinct double-stranded RNA electropherotype pattern in polyacrylamide gels. In Finland, a rotavirus outbreak in ferret kits had a mortality rate that approached 100% [1].

Clinical presentation

Clinical signs of rotavirus infection occur in 2- to 6-week-old ferrets. Soft yellow to green diarrhea is present with associated fecal staining or matted hair on their bodies. Erythema of the anus and perineum also is reported [98]. The disease was prevalent throughout the year at the American commercial farm, with a increased incidence in colder months. Morbidity among primiparous jills was high (up to 90%); the morbidity rate decreased with each gestation to range between 10% to 25% in multiparous jills [98]. The condition could be reproduced in 2- to 3-week-old ferret kits that were inoculated orally with viral preparations that were obtained from diarrheic ferrets; however, mortalities nor histologic lesions were observed in individuals that were infected experimentally [1,98].

Diagnosis

Antemortem diagnosis is difficult. Viral particles may be detected by electron microscopy in clarified, ultracentrifuged fecal suspensions, following negative staining in symptomatic ferrets. In a study, 58% of diarrheic ferrets that were infected naturally were positive for rotavirus particles in their feces. The ferret atypical rotavirus does not react with the Rotazyme test (Abbott Laboratory, Chicago, Illinois), a commercially available enzyme immunoassay [1,98]. The prevalence of this viral infection is unknown in ferrets because of the absence of reliable serologic tests. On postmortem examination, gross lesions are limited to the gastrointestinal tract. Subtle histologic lesions that consist of mild blunting of the tips of the intestinal villi of the small intestine with enterocyte metaplasia to cuboidal cells may be observed.

Treatment/prevention

Secondary bacterial infections may be a significant factor in the severity of the diarrhea. Therefore, antibiotics are indicated and may contribute to accelerated recovery and decreased mortality rates. Additional supportive care, including appropriate fluid therapy and force feeding, are indicated in most cases.

In piglets, colostral antibodies play a key role in the protection against rotavirus [1]. This also may be true for ferrets. Ferret kits acquire most of their passive immune globulins from their mothers by intestinal transmucosal absorption from colostrum and milk; they acquire all of their IgA from their mother's milk [99]. The observed decrease in morbidity from ferrets of primiparous jills to ferrets of multiparous jills may be secondary to build-up of colony immunity with increasing age and exposure to the virus [98].

Oral vaccination of primiparous jills was attempted at a ferret breeding farm; however, this procedure was ineffective. Atypical rotaviruses have not been cultivated successfully in cell culture [1]. The ability to propagate the virus in cell culture in sufficient numbers will play a key role in the development of a vaccine.

Infectious bovine rhinotracheitis virus

The infectious bovine rhinotracheitis virus (IBR) belongs to the family Herpesviridae. There is only one published case report of spontaneous IBR infection in a ferret [100]. The virus was isolated from the liver, the spleen, and the lungs of a clinically normal ferret. Its diet consisted of 5% raw beef by-products; it was hypothesized that virus-laden raw beef was the source of infection. The pathogenesis by which the virus disseminated to the liver, spleen, and lung tissue has not been elucidated. In contrast to naturally-occurring infection, experimental infection of ferrets with IBR virus by intranasal and intraperitoneal inoculations induced acute and chronic respiratory disease [101]. Considering that IBR virus can cause pathology in ferrets, these animals should not be fed raw meat or meat products [1].

Summary

Distemper and rabies vaccination are highly recommended because of the almost invariable fatal outcome of these conditions. Vaccination should constitute an important part of a ferret's preventative medicine program. With the current and anticipated development and licensing of new vaccines, practitioners are invited to gain awareness of the latest vaccine information. Establishment of a practice vaccination protocol with regards to the site of administration of rabies and distemper vaccines is paramount to document any future abnormal tissue reactions.

Influenza is the most common zoonotic disease that is seen in ferrets. Although it generally is benign in most ferrets, veterinarians must take this condition seriously. The characteristic continuous antigenic variation of this virus may lead to more virulent strains; the recent emergence of avian influenza virus outbreaks; and the increased susceptibility of elderly, young, and immunosuppressed individuals.

References

[1] Fox JG, Pearson RC, Gorham JR. Viral diseases. In: Fox JG, editor. Biology and diseases of the ferrets. 2nd edition. Philadelphia: Lippincott Williams & Wilkins; 1998. p. 355–74.

[2] Williams ES, Thorne ET, Appel MJG, et al. Canine distemper in black-footed ferrets (*Mustela nigripes*) from Wyoming. J Wild Dis 1988;24(3):385–98.

[3] Appel MJG, Summers BA. Pathogenicity of morbilliviruses for terrestrial carnivores. Vet Microbiol 1995;44:187–91.

[4] Von Messling V, Springfeld C, Devaux P, et al. A ferret model of canine distemper virus virulence and immunosuppression. J Virol 2003;77(23):12579–91.

[5] Crook E, Gorham JR, McNutt SH. Experimental distemper in mink and ferrets. I. Pathogenesis. Am J Vet Res 1958;19(73):955–7.

[6] Pfeiffer CJ. Gastric hypochlorhydria in ferret distemper. Can J Comp Med Vet Sci 1967; 31:135–8.

[7] Ryland LM. A clinical guide to the pet ferret. Compend Cont Ed 1983;8:25–33.

[8] Ryland LM, Bernard SL, Gorham JR. A clinical guide to the pet ferret. In: Rosenthal KL, editor. Practical exotic animal medicine, the compendium collection. Trenton (NJ): Veterinary Learning System; 1997. p. 122–9.

[9] Davidson M. Canine distemper virus infection in the domestic ferret. Compend Cont Ed Pract Vet 1986;8:448–53.

[10] Stephenson CB, Welter J, Thaker SR, et al. Canine distemper virus (CDV) infection of ferrets as a model for testing Morbillivirus vaccine strategies: NYVAC- and ALVAC-based CDV recombinants protect against symptomatic infection. J Virol 1997;71(2):1506–13.

[11] Rzezutka A, Mizak B. Application of N-PCR for diagnosis of distemper in dogs and fur animals. Vet Mibrobiol 2002;88:95–103.

[12] Blair EM, Chambers MA, King HA. Treating distemper in a young ferret. Vet Med 1998; 93(7):655–8.

[13] Greenacre CB. Incidence of adverse events in ferrets vaccinated with distemper or rabies vaccine:143 cases(1995–2001). J Am Vet Med Assoc 2003;223(5):663–5.

[14] Winsatt J, Jay MT, Innes KE, et al. Serologic evaluation, efficacy, and safety of a commercial modified-live canine distemper vaccine in domestic ferrets. Am J Vet Res 2001;62(5):736–40.

[15] Carpenter JW, Appel MJ, Erickson RC, et al. Fatal vaccine-induced canine distemper virus infection in black-footed ferrets. J Am Vet Med Assoc 1976;169:961–4.

[16] Kauffman CA, Bergman AG, O'Connor RP. Distemper virus infection in ferrets: an animal model of measles-induced immunosuppression. Clin Exp Immunol 1982;47:617–25.

[17] Rosenthal KL. Respiratory diseases. In: Hillyer EV, Quesenberry KE, editors. Ferrets, rabbits and rodents clinical medicine and surgery. Philadelphia: WB Saunders; 1997. p. 77–9.

[18] Quesenberry KE, Orcutt C. Basic approach to veterinary care. In: Quesenberry KE, Carpenter JW, editors. Ferrets, rabbits, and rodents clinical medicine and surgery. 2nd edition. Philadelphia: WB Saunders; 2004. p. 13–24.

[19] Kenyon AJ, Howard E, Buko L. Hyperglobulinemia in ferrets with lymphoproliferative lesions (Aleutian disease). Am J Vet Res 1967;28:1167–72.

[20] Murakami M, Matsuba C, Une Y, et al. Nucleotide sequence and polymerase chain reaction/restriction fragment length polymorphism analyses of Aleutian disease virus in ferrets in Japan. J Vet Diagn Invest 2001;13:337–40.

[21] Porter HC, Porter DD, Larsen AF. Aleutian disease in ferrets. Infect Immun 1982;36: 379–86.

[22] Saifuddin M, Fox JG. Identification of a DNA segment in ferret Aleutian disease virus similar to a hypervariable capsid region in mink Aleutian disease parvovirus. Arch Virol 1996;141:1329–39.

[23] Ohshima K, Shen DT, Henson JB, et al. Comparison of the lesions of Aleutian disease in mink and hypergammaglobulinemia in ferrets. Am J Vet Res 1978;39(4):653–7.

[24] Daoust PY, Hunter DB. Spontaneous Aleutian disease in ferrets. Can Vet J 1978;19:133–5.

[25] Palley LS, Corning BF, Fox JC, et al. Parvovirus-associated syndrome (Aleutian disease) in two ferrets. J Am Vet Med Assoc 1992;201:100–6.

[26] Morrisey JK. Part II. Other diseases. In: Quesenberry KE, Carpenter JW, editors. Ferrets, rabbits, and rodents clinical medicine and surgery. 2nd edition. Philadelphia: WB Saunders; 2004. p. 66–70.

[27] Alexandersen S, Bloom ME. Studies on the sequential development of acute interstitial pneumonia caused by Aleutian disease virus in mink kits. J Virol 1987;61(1):81–6.

[28] Une Y, Wakimoto Y, Nakano Y, et al. Spontaneous Aleutian disease in a ferret. J Vet Med Sci 2000;62:553–5.

[29] Oxenham M. Aleutian disease in the ferret. Vet Rec 1990;126(23):585.

[30] Stewart JD, Rozengurt N. Aleutian disease in the ferret. Vet Rec 1993;133:172.

[31] Welchman D de B, Oxenham M, Done SH. Aleutian disease in domestic ferrets: diagnostic findings and survey results. Vet Rec 1993;132:479–84.

[32] Wolfensohn HHL. Aleutian disease in laboratory ferrets. Vet Rec 1994;134:100.

[33] Oxenham M. Aleutian disease in ferrets. Vet Rec 1992;131(13):296.

[34] Kenyon AJ, Kenyon BJ, Hahn EC. Protides of the Mustelidae immunoresponse of mustelids to Aleutian mink disease virus. Am J Vet Res 1978;39:1011–5.

[35] Bloom ME, Race RE, Wolfinbarger JB. Identification of a non-virion protein of Aleutian disease virus: mink with Aleutian disease have antibody to both virion and nonvirion proteins. J Virol 1982;43:608–16.

[36] Haas L, Lochelt M, Kaaden OR. Detection of Aleutian disease virus DNA in tissues of naturally infected mink. J Gen Virol 1988;69:705–10.

[37] Cheema A, Henson JB, Gorham JR. Aleutian disease of mink: prevention of lesions by immunosuppression. Am J Pathol 1972;55:543–6.

[38] Aasted B, Alexandersen S, Hansen M. Treatment of neonatally Aleutian disease virus (ADV) infected mink kits with gamma-globulin containing antibodies to ADV reduces death rate of mink kits. Acta Vet Scand 1988;29:323–30.

[39] Cho HJ, Greenfield J. Eradication of Aleutian disease of mink by eliminating positive counter immunoelectrophoresis reactors. J Clin Microbiol 1978;7:18–22.

[40] Shen DT, Leendertsen LW, Gorham JR. Evaluation of chemical disinfectants for Aleutian disease virus of mink. Am J Vet Res 1981;42(5):838–40.

[41] Porter DD, Larsen AE, Porter HG. The pathogenesis of Aleutian disease of mink. II. Response of mink to formalin treated diseased tissue and to subsequent challenge with virulent inoculum. Can J Comp Med 1963;27:124–8.

[42] Gorham JR, Evermann JF, Ward A. Detection of coronavirus-like particles from mink. Can J Vet Res 1990;54:383–4.

[43] Have P, Moving V, Svansson V, et al. Coronavirus infection in mink (*Mustela vison*). Serological evidence of infection with a coronavirus related to transmissible gastroenteritis virus and porcine epidemic diarrhea virus. Vet Microbiol 1992;31:1–10.

[44] Williams BH, Kiupel M, West KH, et al. Coronavirus-associated epizootic catarrhal enteritis in ferrets. J Am Vet Med Assoc 2000;217(4):526–30.

[45] Hoefer HL. Gastrointestinal diseases. In: Hiller EV, Quesenberry KE, editors. Ferrets, rabbits and rodents clinical medicine and surgery. 1st edition. Philadelphia: WB Saunders; 1997. p. 26–36.

[46] Doggart L. Viral disease of pet ferrets: part II. Aleutian disease, influenza, and rabies. Vet Technol 1988;8:384–9.

[47] Marois P, Boudreault A, DiFranco F, et al. Response of ferrets and monkeys to intranasal infection with human, equine, and avian influenza viruses. Can J Comp Med 1971;35:71–6.

[48] Shope RE. The infection of ferrets with swine influenza virus. J Exp Med 1934;60:49–61.

[49] Zitzow LA, Rowe T, Morken T, et al. Pathogenesis of avian influenza A (H5N1) viruses in ferrets. J Virol 2002;76(9):4420–9.

[50] Smith W, Andrews DH, Laidlow PO. The virus obtained from influenza patients. Lancet 1933;2:66.

[51] Smith W, Stuart-Harris CH. Influenza infection of man from the ferret. Lancet 1936;2:21.

[52] Smith H, Sweet C. Lessons for human influenza from pathogenicity studies in ferrets. Rev Infec Dis 1998;10:56–75.

[53] Sweet C, Fenton RJ, Price GE. The ferret as an animal model of influenza virus infection. In: Zak O, Sande MA, editors. Handbook of animal models of infection. New York: Academic Press; 1999. p. 989–98.

[54] Basarab O, Smith H. Quantitative studies on the tissue localization of influenza virus in ferrets after intranasal and intravenous or extracordial inoculation. Br J Exp Pathol 1969; 50:612.

[55] Marini RP, Adkins JA, Fox JG. Proven and potential zoonotic diseases of ferrets. J Am Vet Med Assoc 1989;195:990–4.

[56] Collie MH, Rushton DI, Sweet C, et al. Studies of influenza infection in newborn ferrets. J Med Microbiol 1980;13(4):561–71.

[57] Buchman CA, Swarts JD, Seroky JT, et al. Otologic and systemic manifestations of experimental influenza A virus infection in the ferret. Otolaryngol Head Neck Surg 1995; 112(4):572–8.

[58] Rosenthal KL. Respiratory disease. In: Quesenberry KE, Carpenter JW, editors. Ferrets, rabbits and rodents clinical medicine and surgery. 2nd edition. Philadelphia: WB Saunders; 2004. p. 72–8.

[59] Sweet C, Macartney JC, Bird RA, et al. Differential distribution of virus and histological damage in the lower respiratory tract of ferrets infected with influenza viruses of differing virulence. J Gen Virol 1981;54:103–14.

[60] Coates DM, Husseini RH, Rushton DI, et al. The role of lung development in the age-related susceptibility of ferrets to influenza virus. Br J Exp Pathol 1984;65:543.

[61] Sweet O, Jakeman KJ, Rushton I, et al. Role of upper respiratory tract infection in the deaths occurring in neonatal ferrets infected with influenza virus. Microb Pathog 1988;5: 121–5.

[62] Glathe H, Lebhardt A, Hilgenfeld M, et al. Intestinal influenza infection in ferrets (in German). Arch Exp Veterinarmed 1984;38:771–7.

[63] Kang ES, Lee HJ, Boulet J, et al. Potential for hepatic and renal dysfunction during influenza B infection, convalescence, and after induction of secondary viremia. J Exp Pathol 1992;6:133–44.

[64] Rowe T, Cho DS, Bright RA, et al. Neurological manifestations of avian influenza viruses in mammals. Avian Dis 2003;47:1122–6.

[65] De Boer GF, Back W, Osterhaus AD. An ELISA for detection of antibodies against influenza A nucleoprotein in humans and various animal species. Arch Virol 1990;115: 47–61.

[66] Sweet C, Bird RA, Cavanah D, et al. The local origin of the febrile response induced in ferrets during respiratory infection with a virulent influenza virus. Br J Exp Pathol 1979; 60:300.

[67] Chen KS, Bharaj SS, King EC. Induction and relief of nasal congestion in ferrets infected with influenza virus. Int J Exp Pathol 1995;76:55–64.

[68] Husseini RH, Collie MH, Rushton DI, et al. The role of naturally-acquired bacterial infection in influenza-related death in neonatal ferrets. Br J Exp Pathol 1983;64:559–69.

[69] Husseini RH, Sweet C, Overton H, et al. Role of maternal immunity in the protection of newborn ferrets against infection with a virulent influenza virus. Immunology 1984;52(3): 389–94.

[70] Fenton RJ, Bessell C, Spilling CR, et al. The effects of per oral or local aerosol administration of 1-aminoadamantane hydrochloride (amantadine hydrochloride) on influenza infections of the ferret. J Antimicrob Chemother 1977;3(5):463.

[71] Fenton RJ, Morley PJ, Owens IJ, et al. Chemoprophylaxis of influenza A virus infections, with single doses of zanamivir, demonstrates that zanamivir is cleared slowly from the respiratory tract. Antimicrob Agents Chemother 1999;43(1):2642–7.

[72] Herlocher ML, Truscon R, Fenton R, et al. Assessment of development of resistance to antivirals in the ferret model of influenza virus infection. J Infec Dis 2003;188(9):1355–61.

[73] Potter CW, Oxford JS, Shore SL, et al. Immunity to influenza infection in ferrets. I. Response to live and killed virus. Br J Exp Pathol 1972;53(2):153.

[74] Husseini RH, Sweet C, Collie MH, et al. Elevation of nasal viral levels by suppression of fever in ferrets infected with influenza viruses of differing virulence. J Inf Dis 1982;145(4): 520–4.

[75] Paisley JW, Lauer BA. Severe facial injuries to infants due to unprovoked attacks by pet ferrets. JAMA 1988;259:2005.

[76] Diesch SL. Reported human injuries or health threats attributed to wild and exotic animals kept as pets (1971–1981). J Am Vet Med Assoc 1982;18:382.

[77] Applegate JA, Walhout MF. Childhood risks from the ferret. J Emerg Med 1998;16(3): 425–7.

[78] National Association of State Public Health Veterinarians. Compendium of animal rabies control. J Am Vet Assoc 1990;196;36–9.

[79] Jenkins SR, Osterholm MT. Epidemiologists and public health veterinarians issue statement on ferrets. J Am Vet Med Assoc 1994;205:534.

[80] Centers for Disease Control. Viral diseases: ferret rabies. Rabies surveillance, annual summary. Washington, DC: US Department of Health and Human Services; 1986.

[81] Niezgoda M, Briggs DJ, Shaddock J, et al. Pathogenesis of experimentally induced rabies in domestic ferrets. Am J Vet Res 1997;58(11):1327–31.

[82] Blancou J, Aubert JFA, Artois M. Rage expérimentale du furet [Mustela (putorius) furo]. [Experimental rabies in the ferret (Mustela (putorius) furo)]. Revue Méd Vét 1982;133:553. (in French)

[83] Niezgoda M, Briggs DJ, Shaddock J, et al. Viral excretion in domestic ferrets (Mustela putorius furo) inoculated with a raccoon rabies isolate. Am J Vet Res 1998;58(12): 1629–32.

[84] Charlton KM. The pathogenesis of rabies and other lyssaviral infections: recent studies. Curr Top Microbiol Immunol 1994;187:95–119.

[85] Bell JF, Moore GJ. Susceptibility of carnivore to rabies virus administered orally. Am J Epidemiol 1971;93:176.

[86] Förster U. Zur frage der adaptationsfähigkeit von zwei in Mitteleuropa isolierten tollwutvirusstâmmen an eine domestizierte und zwei wildlebende spezies. Ein beitrag zur epidemiologic der tollwut 4. Mitteilung: übertragsversuche an frettchen mit einem nagerisolat. [The adaptability of two rabies virus strains isolated in central Europe to one domesticated and two wild-living species: a contribution to epidemiology of rabies. Part 4: transmission studies on ferret with rodent isolate]. Zentralbl Veterinarmed 1979;26: 29–38. (in German)

[87] Rupprecht CE, Gilbert J, Pitts R, et al. Evaluation of an inactivated rabies virus vaccine in domestic ferrets. J Am Vet Med Assoc 1990;196:1614.

[88] Hoover JP, Baldwin CA, Rupprecht CE. Serologic response of domestic ferrets (Mustela putorius furo) to canine distemper and rabies virus vaccines. J Am Vet Med Assoc 1989; 194(2):234–8.

[89] National Association of State Public Health Veterinarians. Compendium of animal rabies prevention and control, 2003. Morb Mortal Wkly Rep 2003;52(RR-5).

[90] Rabies vaccines licensed in Canada. Canadian Food Inspection Agency, Animal Health and Production Division, Veterinary biologics section. Available at: http://www.inspection.gc.ca/english/anima/vetbio/prod/rabrage.shtml. Accessed October 2004.

[91] Murray J. Vaccine injection-site sarcoma in a ferret. J Am Vet Med Assoc 1998;213(7):955.

[92] Hendrick MJ. Historical review and current knowledge of risk factors involved in feline vaccine-associated sarcomas. J Am Vet Med Assoc 1998;213:1422–3.

[93] Hendrick MJ. Feline vaccine-associated sarcomas. Cancer Invest 1999;17:273–7.

[94] Hendrick MJ, Dunagan CA. Focal necrotizing granulomatous panniculitis associated with subcutaneous injection of rabies vaccine in cats and dogs: 10 cases (1988–1989). J Am Vet Med Assoc 1991;198:304–5.

[95] Carroll EE, Dubielzig RR, Schultz RD. Cats differ from mink and ferrets in their response to commercial vaccines: A histologic comparison of early vaccine reactions. Vet Pathol 2002;39:216–27.

[96] Lincoln J. Another interpretation of ferret's reaction to vaccination. J Am Vet Med Assoc 2003;223(8):1112.

[97] Fearneyhough MG. Rabies postexposure prophylaxis. Vet Clin North Am 2001;31(3): 557–72.

[98] Torres-Medina A. Isolation of an atypical rotavirus causing diarrhea in neonatal ferrets. Lab Animal Sc 1987;37(2):167–71.

[99] Suffin SC, Prince GA, Murk KB, et al. Ontogeny of the ferret humoral response. J Immunol 1979;123:6–9.

[100] Porter DD, Larsen AE, Cox NA. Isolation of infectious bovine rhinotracheitis virus from Mustelidae. J Clin Microbiol 1975;1:112–3.

[101] Smith PC. Experimental infectious bovine rhinotracheitis virus infection of English ferrets (*Mustela putorius furo L*). Am J Vet Res 1978;39(8):1369–72.

ELSEVIER
SAUNDERS

Vet Clin Exot Anim 8 (2005) 161–172

VETERINARY
CLINICS
Exotic Animal Practice

Index

Note: Page numbers of article titles are in **boldface** type.

Antiviral therapy. See *specific species or virus.*

Assays, in diagnostic virology, 7–8. See also *specific assay.*

B

Bass, largemouth, iridovirus in, 80–82

Birds. See *Companion birds; specific species.*

Brain, viral disease of, in ferrets, 144

Breeding programs, for virus control, in amphibians, 56
    in ferrets, 154–155
    in reptiles, 30
    in small mammals, 108, 114–115

Bronchitis virus, infectious, in passerine birds, 90
    in pigeons, 89–90

Budgerigars, adenoviruses in, 86
    European, herpesviruses in, 92
    paramyxovirus-5 in, 98

Bunyaviruses, in reptiles, 43

C

Caliciviruses, in amphibians, 62
    in reptiles, 43–44

Canaries, poxvirus in, 99
    retroviruses in, 101

Canine distemper virus (CDV), in ferrets, 139–143
    clinical signs of, 140
    diagnosis of, 140–141
    epidemiology of, 139
    pathogenesis of, 139–140
    prevention of, 142–143
    treatment of, 142–143

Carp pox, 74

Carp virus, spring viremia of, 69–72

Catfish, viral diseases of, 80

Cell cultures, in diagnostic virology, of pet fish, 71, 73, 75, 78, 80–81

Cellular tropism, influence on viral pathogenesis, 4

Centrarchids, viral diseases of, 80–82

Chicken anemia virus (CAV), 87

Chinchillas, viral diseases of, 116

Chromodacryorrhea, 108

Chronic viral infections, 5

Cichlids, viral diseases of, 75–78

Circoviruses, in companion birds, 87–89

Classification, of viruses, 2–3

Clownfish, lymphocystis in, 69
    viral diseases of, 68–69

Cockatiels, adenoviruses in, 86
    paramyxovirus-3 in, 98

Cockatoos, adenoviruses in, 86
    paramyxovirus-3 in, 98

Columbiformes, paramyxovirus in, 98
    poxvirus in, 99

Companion birds, viral diseases of, **85–105**
    adenoviridae, 85–87
    Amazon tracheitis virus, 92
    circoviridae, 87–89
    coronaviridae, 89–90
    diagnostic testing, 86, 88–90, 94–97, 100–102
    flaviviridae, 90–91
    herpesviridae, 91–93
    herpesvirus associated with papillomatosis, 93
    herpesvirus associated with wartlike skin lesions, 92
    herpesvirus of European budgerigars, 92
    in canaries, 99
    in lovebirds, 99
    in parrots, 86–89, 94–95, 99
    in passerines, 90, 94–97, 100
    in pigeons, 86, 89–90, 98–100
    in psittacines, 86–87, 99–102
    influenza viruses, 93–94
    Newcastle's disease virus, 9, 96–97
    orthomyxoviridae, 93–94
    Pacheco's disease virus, 91
    papillomaviridae, 94–95
    papovaviridae, 94–96
    parakeet herpesvirus, 92
    paramyxoviridae in, 96–98
    polyomaviridae, 95–96
    poxviridae, 98–100
    proventricular dilatation disease, 102
    psittacine beak and feather diseases, 87–89
    reoviridae, 100
    retroviridae, 101
    rhabdoviridae, 101
    severe acute respiratory syndrome, 90
    togaviridae, 101–102
    vaccines for, 92, 97, 99–100
    West Nile virus, 90–91